25 Advances in Polymer Science
Fortschritte der Hochpolymeren-Forschung

Edited by H.-J. Cantow, Freiburg i. Br. · G. Dall'Asta, Cesano Maderno
K. Dušek, Prague · J. D. Ferry, Madison · H. Fujita, Osaka · M. Gordon,
Colchester · W. Kern, Mainz · G. Natta, Milano · S. Okamura, Kyoto
C. G. Overberger, Ann Arbor · T. Saegusa, Kyoto · G. V. Schulz,
Mainz · W. P. Slichter, Murray Hill · J. K. Stille, Fort Collins

With 55 Figures

Springer-Verlag
Berlin Heidelberg New York 1977

Editors

Prof. Dr. HANS-JOACHIM CANTOW, Institut für Makromolekulare Chemie der Universität, Stefan-Meier-Str. 31, 7800 Freiburg i. Br., BRD

Prof. Dr. GINO DALL'ASTA, SNIA VISCOSA – Centro Sperimentale, Cesano Maderno (MI), Italia

Prof. Dr. KAREL DUŠEK, Institute of Macromolecular Chemistry, Czechoslovak Academy of Sciences, 162 06 Prague 616, ČSSR

Prof. Dr. JOHN D. FERRY, Department of Chemistry, The University of Wisconsin, Madison 6, Wisconsin 53706, U.S.A.

Prof. Dr. HIROSHI FUJITA, Osaka University, Department of Polymer Science, Toyonaka, Osaka, Japan

Prof. Dr. MANFRED GORDON, University of Essex, Department of Chemistry, Wivenhoe Park, Colchester C04 3 SQ, England

Prof. Dr. WERNER KERN, Institut für Organische Chemie der Universität, 6500 Mainz, BRD

Prof. Dr. GIULIO NATTA, Istituto di Chimica Industriale del Politecnico, Milano, Italia

Prof. Dr. SEIZO OKAMURA, Department of Polymer Chemistry, Kyoto University, Kyoto, Japan

Prof. Dr. CHARLES G. OVERBERGER, The University of Michigan, Department of Chemistry, Ann Arbor, Michigan 48 104, U.S.A.

Prof. TAKEO SAEGUSA, Kyoto University, Department of Synthetic Chemistry, Faculty of Engineering, Kyoto, Japan

Prof. Dr. GÜNTER VICTOR SCHULZ, Institut für Physikalische Chemie der Universität, 6500 Mainz, BRD

Dr. WILLIAM P. SLICHTER, Bell Telephone Laboratories Incorporated, Chemical Physics Research Department, Murray Hill, New Jersey 07 971, U.S.A.

Prof. Dr. JOHN K. STILLE, Colorado State University, Department of Chemistry, Fort Collins, CO 805 23, U.S.A.

ISBN 3-540-08389-8 Springer-Verlag Berlin Heidelberg New York
ISBN 0-387-08389-8 Springer-Verlag New York Heidelberg Berlin

Library of Congress Catalog Card Number 61-642

This work is subject to copyright. All rights are reserved, whether the whole or part of the material is concerned, specifically those of translation, reprinting, re-use of illustrations, broadcasting, reproduction by photocopying, machine or similar means, and storage in data banks. Under § 54 of the German Copyright Law where copies are made for other than private use, a fee is payable to the publisher, the amount to the fee to be determined by agreement with the publisher.

© by Springer-Verlag Berlin Heidelberg 1977
Printed in Germany

The use of general descrive names, trademarks, etc. in this publication, even if the former are not especially identified, is not to be taken as a sign that such names, as understood by the Trade Marks and Merchandise Marks Act, may accordingly be used freely by anyone.

Typesetting and printing: Schwetzinger Verlagsdruckerei. Bookbinding: Brühlsche Universitätsdruckerei, Lahn-Gießen.
2152/3140 – 543210

The Editors of Advances in Polymer Science extend their warmest felicitations to Dr. Maurice Huggins on the occasion of his eightieth birthday on September 19th, 1977. We are happy to dedicate this 25th Volume to him in recognition of his outstanding contributions to Polymer Science and to other branches of physical science.
He was, of course, quite unaware of our intention when he was persuaded to write a preface to this work, and we hope he will forgive us for presenting this volume to him, and his photograph to our readers, without prior permission.

Preface

The scientific literature has become so immense that it is an impossible task for anyone, by reading the original research reports, to become familiar with the background and current status of more than a very limited field. Even in a narrow area, only a few scientists are sufficiently knowledgable and critical to be able to present the kind of review that will be most useful to others.

To newcomers in a scientific field, who have had neither the time to read nor the background to understand what they read, critical reviews by more experienced scientists are especially important. Such reviews help a newcomer to choose the direction of his own research and give him a firm basis on which to build.

A good review of a field will often give interesting and useful ideas to older scientists, already knowledgable in other fields. Other scientists in the same field as that being reviewed can usually profit from careful study of a review, because the writer and the reader have different backgrounds, different evaluations of the relative importance of previous researches, and different ideas as to what should be done next.

Writing a review, like teaching a course, is also useful to the reviewer. He must study carefully the work of others. He must be critical, but fair, in reporting the results of both his own and others' work.

Reviews, such as those that have been published in *Advances in Polymer Science,* thus play a very important role in the education of scientists and therefore in the progress of science. How important this role is, of course, depends on the choice of the reviewers and the quality of their reviews. In my opinion, the editors of the volumes in this series have just cause to be proud of their record over the past 25 volumes.

The current volume contains four reviews that will, I believe, maintain the high standard of previous years.

Davydov and Krentsel discuss the synthesis and properties of polyconjugated systems: organic polymers containing large sequences of alternating single- and double-bonds. In general, these substances differ from ordinary polymers in their ability to transfer polarization (of the electronic structure) in one region to a corresponding polarization in another, distant, region. This can result in interesting light absorption properties (photosensitivity), chemical properties (crosslinking), electrical conductivity, etc.

We are all familiar with chain polymers containing rings incorporated in the chains [cellulose and starch derivatives, poly(ethylene terephthalate), etc.] or attached to the chains (polystyrene) and with chain polymers produced from ring monomers [poly(ethylene oxide)], but few of us are acquainted with the quite extensive research that has been done on the formation and properties of similar polymers made from compounds containing the furan ring. In the present volume, Gandini ably discusses this subject. In view of the ready availability, from various plant residues, of 2-furaldehyde and the ability of this compound and its simple derivatives to enter into polymerization reactions, it is to be expected that it will not be long before interesting polymers from this source will become common and industrially important.

Polyacrylonitrile is one of the polymers of simple chemical composition that has proven most useful, especially in fibrous form. In certain respects, however, –– with regard to the flexibility, resistance to softening and burning at high temperatures, dyeability, and electrical properties, for example –– the fiber properties can be improved by modifications of the chemical composition and structure. Gabrielyan and Rogovin deal extensively with the methods and results of such modifications.

It is well known that the nucleic acids consist of chain polymers, with each chain corresponding to the general formula

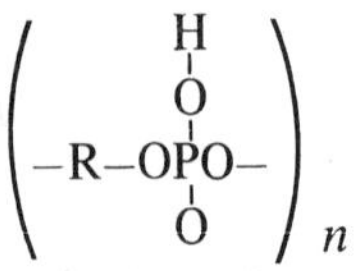

$$\left(-R-O\overset{\overset{\displaystyle H}{\underset{\displaystyle |}{O}}}{\underset{\underset{\displaystyle O}{\displaystyle |}}{P}}O- \right)_n$$

or the polyanion obtained by removing protons from some of the OH groups. Each group here represented as R contains resonating pairs of hydrogen bonding groups that form strong double-hydrogen-bond bridges with similar groups, e. g., in the other chain of a double helix structure.

Since these basic facts became known, a tremendous amount of research has been done on the structures and behaviors of these important substances. There has also been much research on the synthesis and study of other chain polyelectrolytes, containing hydrogen-bond-forming radicals (R) more-or-less like those in the natural nucleic acids. The primary aim of this research is, of course, to relate the behavior of the synthetic materials to the behavior of the natural ones. Okubo and Ise here present an excellent discussion on this research.

135 Northridge Lane
Woodside, California 94062
U.S.A.

Maurice L. Huggins

Contents

Progress in the Chemistry of Polyconjugated Systems
   B. E. Davydov and B. A. Krentsel   1

The Behaviour of Furan Derivatives in Polymerization Reactions
   A. Gandini   47

Chemical Modifications of Fibre Forming Polymers and
Copolymers of Acrylonitrile
   G. A. Gabrielyan and Z. A. Rogovin   97

Synthetic Polyelectrolytes as Models of Nucleic Acids
and Esterases
   T. Okubo and N. Ise   135

Author Index Volumes 1–25   183

Progress in the Chemistry of Polyconjugated Systems

B. E. Davydov and B. A. Krentsel

A. V. Topchiev Institute of Petrochemical Synthesis, Academy of Sciences of the USSR, Moscow, USSR

Table of Contents

1.	**Introduction**	2
2.	**Specific Features of PCS Formation**	4
2.1.	Methods of PCS Formation	4
2.1.1.	Radical Polymerization	5
2.1.2.	Ionic Polymerization	6
2.1.3.	C≡C-Bond Opening *via* Donor-Acceptor Interactions	6
2.1.4.	Polycondensations	7
2.1.5.	Polymeranalogous Reactions	9
2.1.6.	Poly(acrylonitrile) Formation by Intramolecular Polymerization	11
3.	**Effect of Various Factors on the Conjugation Efficiency in PCSs**	17
3.1.	The Length of Conjugation Blocks	17
3.2.	Effects of Conformational Transformations	18
3.3.	Effect of Separation of Conjugated Blocks	21
3.4.	Electronic Excitations and Energy Transfer in PCSs	22
4.	**Chemical Properties of PCSs**	25
4.1.	Reactions with Electrophiles and Nucleophiles	25
4.2.	Thermal Stability. Pyrolysis Reactions	26
4.3.	Polyconjugated Polyelectrolytes	28
4.4.	Photoinitiated Reactions	30
4.5.	Reactions with Dienophiles	31
4.6.	Molecular Complexes and Their Electronic Spectra	31
5.	**Photosensitizing and Catalytic Properties of PCSs**	34
5.1.	Photosensitizing Activity	34
5.2.	Catalytic Activity	36
6.	**References**	38

1. Introduction

The rapid development of macromolecular chemistry has recently brought to light a number of most interesting problems related to the synthesis of polymers with unusual properties. The chemistry of polyconjugated systems undoubtedly occupies a leading place among these problems.

Volkenstein pointed out in his monograph, "Configurational Statistics of Polymer Chains"[1], that a successful synthesis of the (at that time) hypothetic polymer polyvinylene, containing –CH=CH–CH=CH– units in the main chain, could lead to basic transformations in our understanding of the chemical and physicochemical properties of polymers. Within several years such macromolecules were in fact obtained and found to possess outstanding properties. This was the beginning of a new chapter in polymer science, namely, the chemistry of polyconjugated systems (PCSs). This class of substances attracted the attention of scientists in many countries, including the USSR. Within a short space of time, hundreds of compounds of this sort were synthesized. They display a number of unusual properties and open up a fascinating perspective of practical utilization, combining, among other properties, superconductivity and valuable mechanical characteristics.

Subsequent years have demonstrated, however, that the realization of these hopes requires long and systematic studies involving numerous experimental difficulties and new theoretical approaches, differing from those used in the investigation of inorganic semiconductors or organic semiconductors of low molecular weight. Nevertheless, study in depth of the physical and chemical features of PCSs revealed a number of new phenomena which are of general importance for macromolecular chemistry.

The investigation of PCSs was based upon the theoretical prerequisites of the conjugation concept. This concept, developed on the basis of theoretical analysis and experimental studies of the properties of compounds with low molecular weights, has played an essential role in the progress of our understanding of the nature of the chemical bond, structure, and the reactivity of substances.

It is known that the overlap of wave functions of π-electrons of conjugated bonds gives rise to a common multicenter system of electrons delocalized along the

---

[a] Since the specific properties of PCSs result from the overlap of the π-orbitals of multiple bonds, which reach their maximum at the coplanar arrangement of the units in the polyene chain, the perturbation of such an arrangement should lead to a partial or complete loss of the specific properties of a PCS. In particular, since the conjugation energy is proportional to $\cos \alpha$, α being the angle of deviation of fragments of the chain from the coplanar arrangement, an orthogonal structure of the conjugated blocks should result in an independent behavior of the fragments. Therefore, the question which factors perturb the coplanarity of the main polyene chain and to what extent the properties of the polymers are affected, is of fundamental importance.

In this context it is expedient to introduce a "conjugation efficiency" concept to characterize the degree of delocalization of π-electrons and, hence, to understand the phenomena which are typical of PCSs. The conjugation efficiency shows to what number of coplanar, *i.e.*, ideally conjugated, structural fragments corresponds the length of a conjugated block in a real polymer (*e.g.*, according to its absorption or luminescence spectra).

conjugation chain[a]. This leads to an increased mobility of the electrons, to a narrowing of the gap between their energy levels, to a drop in the ionization potential together with an increase of electron affinity, to a rise in polarizability, and to a decrease of the internal energy of the system. All this should be manifested in specific chemical, physicochemical, and physical properties of PCSs and also in specific features of the processes resulting in such systems. The formation and increase in length of the conjugated chain should favor greater susceptibility of a PCS to photoinduced reactions, emergence of photoelectric sensitivity and photoinduced paramagnetism, as well as semiconductive, catalytic, and other specific properties.

An increase in polarizability of the π-electron system in a PCS is expected to be followed by an easy transfer of the chemical behavior along the conjugation chain and by a decrease in the internal energy of the PCS, thus enhancing the thermal stability of the polymers and the rigidity of the main chain.

These considerations lead to the assumption that the practical aspects of the problem lie in the possibility of obtaining PCS-based thermally resistant materials, catalysts for some chemical reactions, antioxidants, stabilizers, photochromic substances, and materials combining valuable mechanical properties with special electrical (particularly semiconductive) properties.

Many processes in living organisms are closely linked to energy transfer and to charge transfer complexes. Therefore, studies of the properties of PCSs are important in solving certain problems of bioenergetics, enzymatic catalysis, photoinduced carcinogenesis, etc.

This review tries to analyze and evaluate the new phenomena disclosed in this fresh field of polymer chemistry and to outline directions of its further development. The properties of polymeric materials and the methods of their synthesis will be considered only so far as is required for the fulfilment of that task. The semiconductive properties of PCSs will not be discussed in detail, since this problem is dealt with in a number of special reviews[2-8] and monographs[9-11] including those written with the participation of the present authors[4,8,10]. Apart from the literature data, the results obtained during the last 15 years by a research team headed by the authors will be considered. The studies have been carried out mainly in the following directions:

1. Specific features of the processes of PCS formation and new reactions of C≡C bond opening.
2. Factors determining the efficiency of conjugation, and the effect of these factors on the properties of PCSs.
3. Transfer of electronic excitation energy in PCSs.
4. Photochemical transformations of PCSs.
5. Chemical properties of PCSs.
6. Donor-acceptor interactions in PCSs.
7. Catalytic and photosensitizing properties of PCSs.

2. Specific Features of PCS Formation

2.1. Methods of PCS Formation

Polymerization and polycondensation processes together with transformations of polymeric chains have been used for the preparation of PCSs. Polymerization processes include the opening of C≡C or C≡N-bonds and the opening of carbo- and heterocycles.

Thus, Natta[12] and other researchers[13,14] polymerized acetylene, using catalysts of the Ziegler-Natta type. In addition alkali borohydrides in combination with nickel halides[15,16], radiation initiation[17], and other methods[18] have been used for the polymerization of acetylene. Acetylene derivatives were also used to obtain PCSs, e.g., ethynylbenzene (phenylacetylene)[19-24], hexyne[25-28], ethynylcyclohexane (cyclohexylacetylene)[27,28], 1-propyne and 1-butyne (methyl- and ethylacetylene), 3-propyne-1-ol (propargyl alcohol)[30], propionitrile[31], acetylenedicarbonitrile (dicyanoacetylene)[29,32], and a number of other monomers[33-35].

Besides, polymerization processes were used to obtain PCSs with alternating C=N-bonds in the main chain. Thus, radiation polymerization of methacrylonitrile[54] gives rise to a ladder polymer of the following type:

(23)

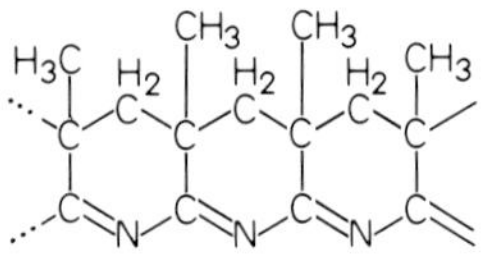

Radiation polymerization of acrylonitrile leads to a PCS with conjugated C=N-bonds and remaining unsaturation in the side chain substituents[55]. Polymerization of stoichiometric complexes of nitriles (propionitrile, acetonitrile, capronitrile) in the presence of metal halides, e.g., $ZnCl_2$, proved to be very effective in making systems with polyconjugated C=N-bonds[56,57]. The same principle was applied for the opening of a number of heterocycles (e.g., pyridine, quinoline)[58-63]. The fact that this process proceeds as an autocatalytic reaction is an interesting feature. The catalytic activity of a deliberately introduced polymer depends considerably on the structural correspondence between this polymer and the polymer obtained in the course of the reaction[59,63]. This, in particular, reflects polymer-monomer interaction which is, apparently, typical of PCSs.

The following examples are typical of polymerization processes giving rise to PCSs: polymerizations of acetylene and diphenylacetylene[36-41] catalyzed by a system of ionic coordination type obtained by reacting vanadium acetylacetonate with Et_3Al (cf.[38]), thermal and radiation polymerization of propynoic acid and its salts[40,41-45], free ionic polymerization of phenylacetylene[46-48], and, finally, the polymerization of NN-dimethyl-2-propynylamine by the opening of the C≡C-bond and simultaneous donor-acceptor interaction of the amino function with organic electron acceptors[49]. The specific features of these processes will be treated in the following chapters. A detailed study of the kinetics of polymerization of acetylene,

1,4-diphenyl-1,3-butadiyne (diphenyldiacetylene), and propynoic acid led to the conclusion that the reactivity of the active centers drops during the growth of the polymeric chain. This is one of the most important characteristics of the polymerization processes resulting in the formation of PCSs.

2.1.1. Radical Polymerizations

In the process of radical polymerization a monomolecular short stop of the kinetic chain arises from the delocalization of the unpaired electron along the conjugated chain and from the competition of the developing polyconjugated system with the monomer for the delivery of π-electrons to the d-orbitals of a transition metal catalyst in the ionic coordination process. Such a deactivation of the active center may also be due to an interaction with the conjugated bonds of systems which have already been formed.

The increased rigidity of the polymer molecules (resulting, as was mentioned above, from the delocalization of π-electrons along the conjugation chain) implies that PCSs may consist of stable *cisoid* and *transoid* conformers. In other words, the rigidity of the polyene chain gives rise to the stereoisomerism peculiar to PCSs. This makes it possible to obtain isomeric PCSs with side groups on one side or on both sides of the main chain. This possibility was realized in polymerizing propynoic acid. Thus, radiation polymerization of this monomer in solid phase (at a temperature of $-77°$ C) results in a polymer (I) with a prevalent *trans-transoid* configuration of the main chain, having carboxyl groups on one side.

(24)

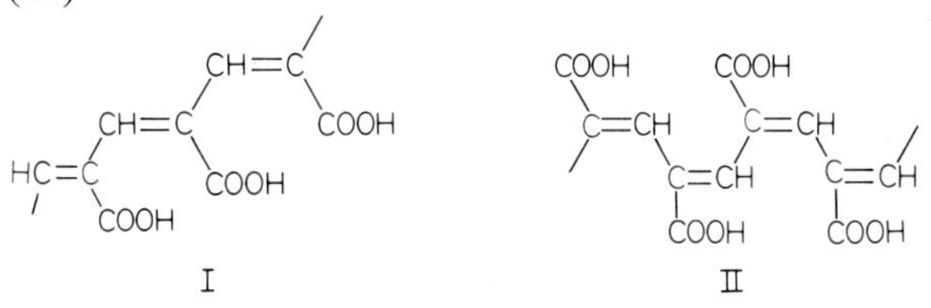

Alternatively, liquid phase polymerization (in bulk monomer at a temperature of $20°$ C) furnishes an isomer (II) characterized by a *cis-transoid* (or *trans-cisoid*) configuration of the main chain, with carboxyl groups located on both sides of it. These isomers will be shown later to differ in chemical and physicochemical properties.

We have shown that polymerization of propynoates always proceeds with the breakdown of the crystalline lattice, the conversion degree depending on the lattice strength. Only propynoates having cations with ionic radii from 0.92 to 0.95 Å show highest polymerization rate and conversion. This allows the suggestion with these particular ionic radii the geometric arrangement of the monomer in the crystal shows the best correspondence to the mutual location of structural fragments in a polymer[42, 43]. It is noteworthy that simultaneous condensation of methyl propiolate and metallic magnesium in molecular beams, followed by thawing, results in a rapid polymerization in the phase transition point of the monomer ($-85°$ C), accompanied by intensive luminescence[41] ($\lambda_{max} = 560$ nm).

2.1.2. Ionic Polymerization

Another interesting feature of a polymerization process producing a PCS was revealed in the free ionic polymerization of phenylacetylene induced by butyllithium in heptane/hexamethylphosphortriamide (30/70). In this solvent mixture, 60% of butyllithium is dissociated into free ions. A detailed study of the kinetics of this process showed[46, 47] that the reaction proceeds on living chains, *i.e.*, no deactivation of the free amine, bound to the conjugated system, occurs. Nevertheless, the polymerization reaction stops in spite of the fact that the reaction medium contains up to 20% of unreacted monomer. The introduction of an additional quantity of the monomer into the reaction system leads to its polymerization until the polymer/monomer ratio again reaches 3/2. A rise in temperature shifts the equilibrium toward polymer formation, thus indicating that this effect is not a consequence of the polymerization-depolymerization equilibrium in the system. All these facts are evidence that, firstly, deactivation of active centers does not occur during polymerization of living chains and, secondly, interaction between the PCS formed and the unreacted monomer takes place, resulting in deactivation of the latter.

The nature of this interaction is not yet clear, but there is no doubt that this is a manifestation of a polymer-monomer interaction, typical of PCSs. The process of polymerization of phenylacetylene on free ions is characterized by the 6th order in initiator and monomer and has an activation energy of ≈ 25 kJ/mol (6 kcal/mol).

2.1.3. C≡C-Bond Opening *via* Donor-Acceptor Interactions

Donor-acceptor interaction is known in a number of cases[51] to favor the reactivity of a monomer. The question, to what extent can this principle be used in obtaining PCSs by opening C≡C-bonds, arose from a detailed study of this problem by the authors, experimenting with a number of heterocycles[52, 53]. We shall consider now our new approach to the C≡C-bond opening to be based upon monomer activation, resulting from its interaction with electron acceptors[49, 50].

The feasibility of this process was studied in the case of the polymerization of *N,N*-dimethyl-2-propynylamine (diethylpropargylamine, DEPA) which cannot be polymerized by usual methods. This monomer was found to interact with such electron acceptors as ethylene tetracarbonitrile (tetracyanoethylene), trinitrobenzene, etc., giving rise to charge transfer complexes capable of spontaneous polymerization (at 40–60° C) without any initiator[49, 50]. A detailed study of the process of complex formation, together with the polymerization reaction has shown that the interaction of DEPA with electron acceptors results in a complex of varying composition (mainly 1:1). Complexes with partial and complete charge transfer are formed in the process, the proportion of the complexes depending on the electron affinity of the acceptor. However, in the presence of such a strong acceptor as tetracyanoethylene, the amount of molecules in the state of complete electron transfer does not exceed 1–2% of the total content of the complex, judging from ESR spectra. Weak acceptors (dinitrobenzene) form charge transfer complexes only of a partial type. Polymerization of complex-bound DEPA proceeds spontaneously at 40–60° C and also photochemically at lower temperatures. The introduction of radical initiators does not affect the polymerization rate, but cationic initiators accelerate the process. The polymerization involves the complex with 1:1 composition including partial and complete charge transfer entities, since the polymer yield exceeds the content of complete charge transfer complex by several orders of magnitude. As the rate of the process involving different acceptors depends on the content of complete charge transfer complexes, there are good reasons to believe that these complexes are responsible for the initiation of polymerization. This means that the process as a whole is initiated by the cationic function of an initiator forming a cation-radical[49, 50].

It can be expected that the results obtained in the polymerization involving C≡C-bond opening and PCS formation caused by a donor-acceptor interaction are of a general nature and may be extended to a rather broad range of monomers.

2.1.4. Polycondensations

Polycondensation processes possess outstanding capabilities, for they allow the complex of properties of a PCSs to vary considerably, depending on the change of the nature of the main chain or on the character of the functional groups. Only through polycondensation reactions it has been possible to obtain a large number of PCSs of different chemical natures. Berlin and co-workers[41] made an attempt to systematize the types of polycondensation processes giving rise to PCSs.

Oxidative polycondensations involving the elimination of hydrogen (or other atoms) as a result of the interaction of monomers with oxygen or other oxidizing agents, belong to these processes. Oxidative polycondensations make it possible to obtain PCSs with carbon-chains or chains with heteroatoms such as polyacetylenes[64–66], polyphenylenes, aminoquinones[67, 68], polyarylenes[69–71], polyanilines[72–75], polyazoarylenes[76, 77], poly(phenylene oxides)[77–79], and "vat" polymers[80–82]. In general terms, oxidative dehydropolycondensation may be expressed as shown in the following scheme:

(25)

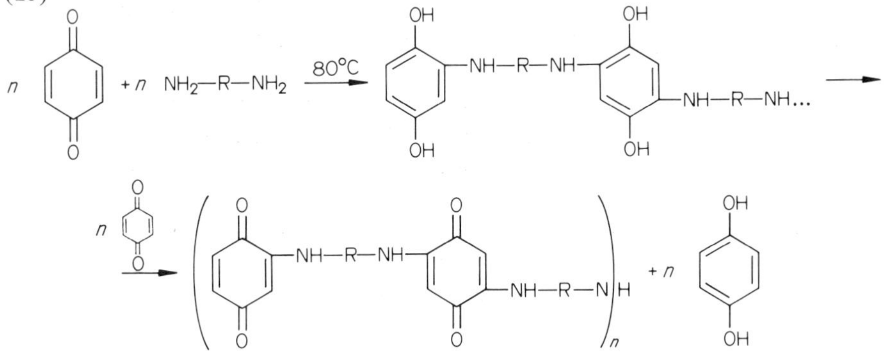

Another group of reactions can be classified as intermolecular dehalogenations or dehydrohalogenations, based on Wurtz-, Fittig-, Ullmann-, Grignard-, or Friedel-Crafts reactions. Thus, polyphenylenes may be obtained by polycondensations of p-dichlorobenzene in the presence of Na/K alloy[83–85].

A different version of these reactions has been used by other authors[86-89]. The Ullmann reaction was applied for the preparation of alkyl substituted polyphenylenes[90-92]. Dehydrohalogenation has been shown to give rise to PCSs with heteroatoms in the main chain[93-98]. Reactions based on the splitting of diazogroups[99-100], including the interaction of bisdiazonium compounds with ferrocene and quinones, as well as the synthesis of polyazo compounds[103, 104] offer interesting opportunities for the preparation of PCSs, particularly those containing various functional groups. The so-called onium polymerization also follows a polycondensation mechanism and results in polymeric quaternary ammonium bases[105-107]. In order to create PCSs with metal atoms, some authors[108-110] used a polycoordination reaction comprising the formation of the main polymer chain by combination of covalent and coordination bonds. This reaction makes it possible to prepare polymers containing porphyrine cycles in a conjugated chain[111-114] (e.g., polyphthalocyanines[111-113], and polymeric complexes of tetracyanoethylene and its analogues[108, 113, 114].

There is great potential for obtaining PCSs by polycondensation processes involving carbonyl groups. Thus, hydrazine and diamines as comonomers give rise to polyazines[40, 115-122] and polymeric Schiff bases[40, 121-131], respectively. On the other hand, polyenes[132, 133] and PCSs containing vinyl bonds and pyridine cycles in the main chain[134, 135] are obtained from dicarbonyl compounds by aldol-crotonaldehyde condensation. The interaction between triphenylphosphinalkylidenes and aromatic and polyenic dialdehydes (Wittig reaction) yields PCSs containing acyclic and aromatic conjugations[121, 122, 136, 137].

Cyclopolycondensation and cyclopolyaddition reactions acquired importance in obtaining PCSs. Thus, the polycondensation of oxamide[138, 139] has furnished a polymer with condensed pyrazine cycles in the main chain. Poly(benzothiazole)s[144], poly(benzimidazole)s and their analogues[140-144], poly(pyromellitimide)s[145], poly(imidazopyrrolone)s[146-149], and poly(imidazopyrone)s[150] have been obtained in the same type of reactions. These reactions are becoming important for making materials of specified structures by adequate processabilities due to the possibility of performing stepwise processes, i.e., by preparation of soluble prepolymers in the first stage with subsequent cyclization to PCSs. Therefore, the second stage can be regarded as a process of PCS preparation by polymeranalogous reactions of polymers obtained in the first stage.

It is known that cyclopolyadditions consist in the interaction of compounds containing a polarized or resonance-stabilized dipole (a 1,3-dipole) with compounds having a system of multiple bonds (dipolarophiles). Applying this principle to the synthesis of polymers, the monomers must have at least two dipole groups and two groups of multiple bonds, respectively. Reactions of this type (known as 1,3-dipolar cyclopolyadditions[41]) were used for the synthesis of PCSs, poly(arylpyrazole)s in particular, by interaction of bis(diazo)alkanes (1,3-dipoles) and diynes (dipolarophiles)[152, 153]. The reaction is formulated as follows:

(26)

$$n\ N{\equiv}N{-}\overset{\oplus\ \ \ \ominus}{CH}{-}\!\!\left\langle\;\right\rangle\!\!{-}\overset{\ominus\ \ \oplus}{CH}{-}N{\equiv}N + n\ HC{\equiv}C{-}\!\!\left\langle\;\right\rangle\!\!{-}C{\equiv}CH \xrightarrow{20°C}$$

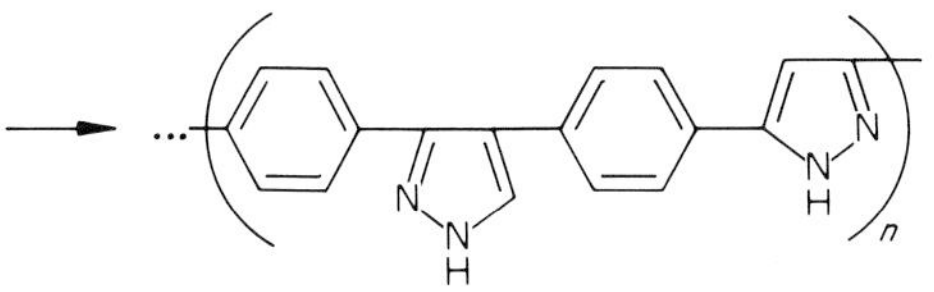

Some researchers have elaborated synthesis of PCSs employing bis(nitrile oxide)s as 1,3-dipoles and diynes, dinitriles, and a number of other compounds as dipolarophiles[154–156].

The thermolytic polycondensation of aromatic hydrocarbons[157], the low-temperature polycondensation of carbohydrates[158], and the so-called combined polyreactions[159–162] were also used for obtaining PCSs. The latter reactions consist in a single process involving the formation of respective monomers and the PCS synthesis. Thus, dehydrobromination of dibromopropionitrile in the presence of CaO and $CuCl_2$ at high temperatures leads to polymers containing cyanovinylidene units and condensed pyridine cycles. Preparation of poly(vinylene)s by reacting chloroform with metallic lithium[162] belongs to the same group of processes. A detailed study of the condensation of hydrazine and diamines with dicarbonyl compounds, resulting in polyazines and poly(schiff base)s, has shown that the process has the following features:

1. The yield and the molecular weight of the polymers rapidly reach the maximum values (up to 6000–8000) irrespective of the concentration of reagents or the process temperature.
2. The reaction stops regardless of a large amount of unreacted amine or hydrazine remaining in the system (even in the case of withdrawal of the evolving water from the system).
3. Polymers obtained have a low molecular weight and contain carbonyl end groups, even in the case of initial systems with a 1:1 mole ratio of diamine to dicarbonyl compound.

This leads to the conclusion that polycondensation processes also feature the deactivation of end groups in PCSs. Mainly carbonyl groups undergo such a deactivation, thus indicating an elevated nucleophilic activity of PCSs. The discovery of this feature made it possible to approach the synthesis of PCSs with higher molecular weights. This is done by polycondensation of monomers containing both carbonyl and amine groups in the molecule. Thus, polycondensation of monohydrazones of diketones allowed the production of polyazines with molecular weights approaching 15000–20000.

2.1.5. Polymeranalogous Reactions

PCSs can also be obtained by polymeranalogous transformations of already available polymers. Splitting of low-molecular vinyl polymers is one of the reactions of this type. These polymers form the following series[163] according to their capability of elimination of low-molecular substances: vinyl chloride copolymers > chlorinated PVCs > poly(vinyl ether)s > poly(vinyl alcohol) > poly(vinyl acetate). The capability of eliminating hydrogen halogenide decreases when passing from iodine- to bromine-containing polymers[164, 165]. PVC dehydrochlorination has been studied in more detail[163, 166–170]. Alcoholates[163, 171] or phenolates[172] of alkali metals, sodium or potassium amides[163, 170, 173], amines[174–175], heterocyclic nitrogen-containing compounds[176], alkali solutions[169, 177], and a number of other substances[177, 179] have been used as dehydrochlorinating agents. According to calculations[168] and experiments[167], the maximum dehydrochlorination degree of PVC reaches 86%. However, there exists information on higher degrees[171]. PCSs were also obtained by dehydrochlorination of chlorinated PVC[180–182], poly(vinylidene chlorofluoride)[183], and a variety of other similar polymers[184]. Exhaustive dehydrochlorination of poly(vinylidene chloride) gave polymers with conjugated triple bonds, poly(ynes) (carbines)[169]. PCSs can be produced by elimination of methoxy groups from poly(methyl vinyl ether) with the help of phenyllithium[180] or by treatment of poly(vinyl alcohol) and poly(vinyl acetate) with thionyl chloride, or orthophosphoric or sulphuric aicd[186]. The processes of thermal elimination of hydrogen halogenide reveals the peculiarity of not following the stochastic law. The elimination of hydrogen halogenide accompanied by double bond formation, occurs initially at the branching sites of the main chain or at linkages adjacent to the unsaturated end groups. The formation of a double bond is followed by an increased mobility of the halogen atom next to the allylic carbon since, e.g., the C–Cl bond energy in allyl chloride (242.8 kJ/mol (58 kcal/mol)) is considerably lower than that in ethyl chloride [335 kJ/mol (80 kcal/mol)][187]. Therefore, the rate constant of elimination of successive halogen atoms increases considerably. This is also favored by the fact that proton-donating additives, e.g., halogen acids, catalyze the thermal HCl elimination[188–192], since the attack of protons at labile chlorine atoms in α-position relative to the arising system of conjugated bonds, weakens its C–Cl bonds to a high degree. Finally, the emerging conjugated system can catalyze the process for reasons which will be discussed later. Therefore, PCS formation *via* thermal elimination of hydrogen halogenides is of an autocatalytic nature. This is particularly the case at temperatures > 220–230° C. To account for the act of primary elimination of hydrogen halogenide, the concepts of free radical[193], molecular[194, 195], and ionic[196] mechanisms of the process have been used.

Processes of PCS preparation by dehydrogenation of saturated, and some unsaturated, polymers are of great interest. Organic and inorganic oxidizing agents with high oxidation potentials, such as quinone, sulfur, N-bromosuccinimide[197], potassium perchlorate[198], chloroanil[199–201], etc., have been used as dehydrogenation agents. Thus, the dehydrogenation of 1,3-polycyclohexadiene gave polyphenylene[197], and the decarboxylation of poly(dimethyl 2,2'-dimethylenepimelate) followed by dehydrogenation resulted in polybenzyl[198]. Readily proceeding dehydrogenations of polybutadiene, polyisoprene, and polystyrene by chloranil[199–201] are due to the increased mobility of the methylene hydrogen atom located in α-position of the double bond. Dehydrogenation of polymers with *trans* configuration occurs more rapidly than that of the *cis* isomers. Polyisoprene undergoes dehydrogenation somewhat easier than polybutadiene. Dehydrogenation of saturated polymers takes place at higher temperatures and is characterized by higher activation energies [up to 105–113 kJ/mol (25–27 kcal/mol)]. Polystyrene is specific in that at small degrees of dehydrogenation no conjugation sections appear with the number of units exceeding three or four. High degrees of dehydrogenation[200–201] lead to dark-coloured insoluble polymers similar to polyacetylene and its homologues. Polyacenaphthylene[200-203] was also subject to dehydrogenation by interaction with chloranil or bromoanil, and to thermal dehydrogenation giving rise to PCSs. Another method of producing PCSs consists in the polymeranalogous transformations of polymers containing carbonyl groups and mobile hydrogen atoms, the transformations being of aldol-crotonaldehyde condensation type. Thus, PCSs are obtained by polymeranalogous transformations of poly(methyl vinyl ketone)[204, 205] and copolymers of methyl vinyl ketone and acrylonitrile[205]. In the latter case conjugated C=N bonds arise at a high rate. Spontaneous dehydrogenation and dehydrochlorination of poly(2-chlorovinyl methyl ketone)[206, 207] leads to a ladder PCS. This is one of the systems which exemplifies the catalytic activity of PCSs in a number of chemical reactions. It has been proposed to obtain PCSs by thermal dehydration of poly(vinyl alcohol)[188, 208, 209], elimination of acetic acid from poly(vinyl acetate)[210], and, finally, exposure of polyethylene, poly(vinyl chloride), poly(methyl methacrylate), poly(vinyl acetate), and a variety of other polymers to radiation[211] or radiation combined with heat[212-216]. Double bonds arising in the initial stage serve as centers initiating chain dehydrochlorination of poly(vinyl chloride)[217].

2.1.6. Poly(acrylonitrile) Formation by Intramolecular Polymerization

Polymeranalogous reactions considered above may be referred to as "intramolecular condensation transformations" since they are accompanied by elimination of low-molecular products. On the other hand, PCSs can be obtained via polymeranalogous transformations, principally "intramolecular polymerization reactions". Thermal and chemical cyclization of poly(acrilonitrile) (PAN) is an example of processes of this type. It was demonstrated by a number of researchers[216-225] that thermal transformations of PAN follow the scheme:

(27)

Phenols, alkali[225], and some other substances[226-228] serve as initiators of these reactions. In the case of poly(acrylonitrile) this role is played by a tertiary hydrogen which is activated by the C≡N group.

Similar transformations have also been performed on model compounds[229-230]. Specific features of polymeranalogous transformations giving PCSs are due to the fact that, on the one hand, PCS formation requires certain changes in the conformation of the polymeric chain, allowing the formation of coplanar blocks. On the other hand, the mutual position of side groups participating in the reaction determines the kinetic and thermodynamic parameters of the process. Along with this, a substantial contribution to the process as a whole is made by intra- and intermolecular interactions such as association, autocatalysis, energy transfer, donor-acceptor interaction, etc. These problems were investigated by the authors[40, 41, 231-244], considering PAN as an example. We shall discuss this process in more detail.

IR spectra of PAN reveal a new band at 1600 cm^{-1} appearing with increasing intensity during the thermal transformation. This band proves the formation of a system of conjugated C=N bonds. The intensity of the band at 2240 cm^{-1} decreases because of the consumption of C≡N groups during the reaction. Generation of methyleneimino groups, resulting from migration of the tertiary hydrogen atom, is the initiating stage of the process. This is demonstrated by the absorption band at 3200–3400 cm^{-1}, corresponding to stretching vibrations of the hydrogen in the C=N–H group[233–237]. Electronic spectra also reveal new absorption bands[234]. Conformational transformations in the polymeric chains with increased rigidity of the chains are indicated by viscosity characteristics of solutions and by the temperature dependence of the square of the second moment (H^2) in the NMR spectra of the polymers[40, 231, 209]. These data are presented in Fig. 1. The reaction under consideration is in fact, as was pointed out earlier, a process of polymerization of

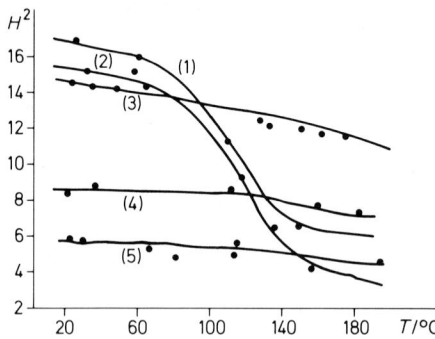

Fig. 1. Dependence of the square of the second moment (H^2) on the temperature for thermally treated PAN (heated at 200 °C). (1) : 3 h; (2) : 5 h; (3) : 15 h; (4) : 40 h; (5) : 150 h

nitrile groups attached to a polyolefinic chain. In a sense, this process should be considered as a matrix polymerization of side groups, the main polymeric chain playing the role of the matrix. This determines the kinetics of the process as a whole and the structure of the products obtained. Examination of molecular models[238] shows that a mutual arrangement of the nitrile groups in *isotactic* PAN sequences is more favorable for an intramolecular cyclization with PCS formation than in *syndiotactic* ones. In order to elucidate the role of the configuration, the thermal transformation processes of PAN have been studied[238–241] using two samples, PAN-r, obtained by radical polymerization of acrylonitrile in solution, and PAN-c, produced by radical polymerization of acrylonitrile in urea clathrate complexes. According to the NMR spectra, the ratios of *isotactic to syndiotactic* diads in PAN-r and PAN-c equals, respectively, 1:1 and 3:1. The conclusions about the kinetics of PCS formation and the character of change of effective conjugation as the process runs, were based on the absorption spectra in IR and UV regions and on the luminescence spectra. From IR spectra we infer that in both cases, in solution and in solid phase, the overall rate of thermal transformation (at 150 °C) of PAN-c is considerably higher than that of PAN-r. This is due to a high growth rate of conjugation blocks in PAN-c, the initiation rate being practically similar to that of PAN-r[238, 239]. The higher rate of formation of the system of conjugated bonds in the polymer with an increased content of *isotactic* sequences (PAN-c) also turns out in the UV spectra, as is shown by Figs. 2 and 3. Moreover, the data in Figs. 2 and 3 imply that the content of longer conjugation sections is considerably greater in thermally transformed products of PAN-c than in thermally transformed PAN-r products. After a heating time of 20 h at 150 °C, the transformed products of PAN-c have lost their solubility completely, whereas the transformed products of PAN-r do not precipitate from the solution even after a heating time of 80–90 h at this temperature. Figure 4 illustrates the effect of PAN microtacticity on the overall kinetics of accumulation of sections with conjugation systems in the polymer. Figure 5 depicts the character of changes in the length of conjugation blocks at early transformation stages. It can be seen that at 150 °C the polymer with an elevated content of *isotactic* units (PAN-c)

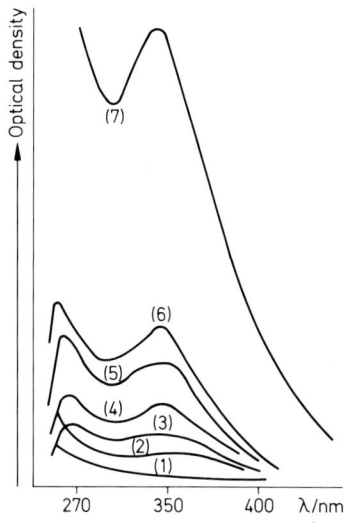

Fig. 2. UV absorption spectra of starting PAN-c (1) and at different steps of thermal treatment in dimethylformamide (DMF) solution at 150 °C. (2) : 1 h; (3) : 2 h; (4) : 3 h; (5) : 6 h; (6) : 10 h; (7) : 15 h

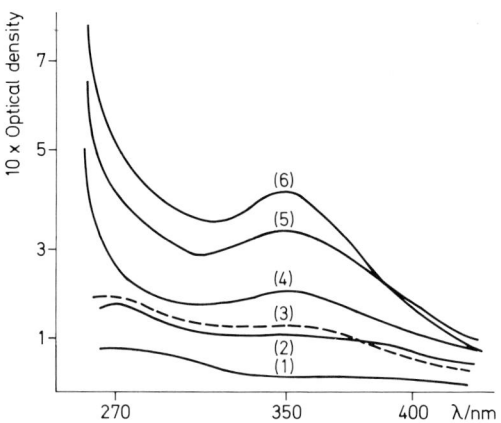

Fig. 3. UV absorption spectra of starting PAN-t (1) and at different steps of thermal treatment in DMF solution at 150 °C. (2) : 3 h; (3) : 6 h; (4) : 10 h; (5) 15 h; (6) : 20 h

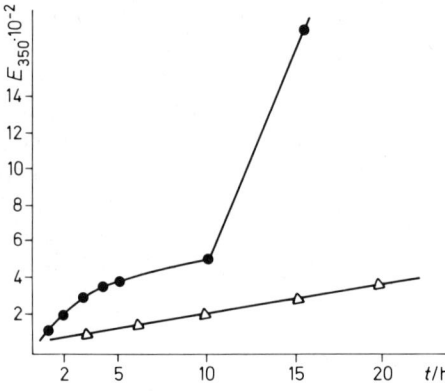

Fig. 4. Variation of the absorption of the coefficient (ϵ) in the UV spectra at $\lambda = 350$ nm for PAN-c and PAN-r with respect to the heating time (t). (●) : PAN-c; (△) : PAN-r

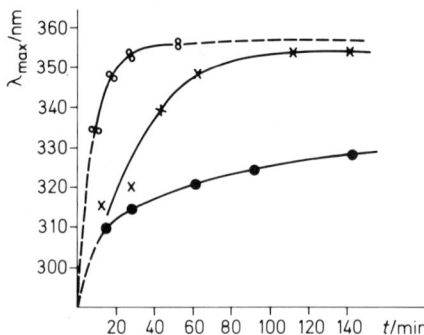

Fig. 5. Variation of λ_{max} in the UV absorption spectra of thermally treated films with time (t). (○) : PAN-r; (×) : PAN-c (at 150 °C); (●) : PAN-r and PAN-c (at 200 °C)

reveals a more sharp increase of length of conjugation blocks. (At higher temperatures the differences in structure of the polymer, in fact, level off.) The data in Fig. 5 reflect, in essence, the kinetics of growth of effective conjugation length in the process of thermal transformation of PAN. The most useful information for the quantitative evaluation of the process can be obtained on the basis of the luminescence spectra at early stages. Luminescence spectra of the starting PAN contain a band at 440 nm related to chromophoric groups of β-ketonitrile type, present in the polymer[245, 265]. At 150–200 °C, within 1–2 min, a band, having a maximum at 460 nm, appears, arising from the formation of a system of conjugated bonds. With heating, the maximum of this band shifts towards longer wavelengths, thus indicating the growth of the effective conjugation chain. The kinetic curves of the process (Fig. 6) demonstrate that the conjugation efficiency increases in PAN-c at a higher rate and reaches higher limiting values than in PAN-r. Thermal transformation of PAN in solution does not change the general character of the process. Figure 7 presents the linearity of the PAN-c transformation kinetics. The parameter $P = (\lambda_t - \lambda_{init})/(\lambda_\infty - \lambda_{init})$ (where λ_∞, λ_t, and λ_{init} refer to the ultimate position of the maximum in the luminescence spectrum, to the current, and to the initial position, respectively) corresponds to the degree of transformation of nitrile groups in the process of the development of effective conjugation.

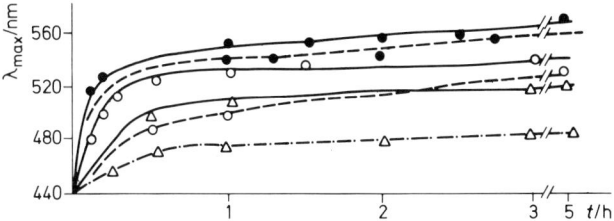

Fig. 6. Kinetic curves of the shift of λ_{max} in the luminescence spectra of PAN samples thermally treated in solid phase. (●) : at 200 °C; (○) : at 180 °C; (△) : 150 °C. Dotted lines refer to PAN-r, solid lines refer to PAN-c

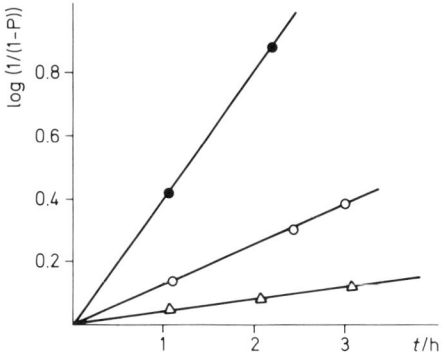

Fig. 7. Time dependence of $\log 1/(1-P)$ for PAN-c samples heated in DMF solution. $P = (\lambda_t - \lambda_{init})/(\lambda'_\infty - \lambda_{init})$; λ_∞, λ_t, and λ_{init} refer to the ultimate position of the maximum in the luminescence spectrum, to the current and to the initial position, respectively. (●) : at 180 °C; (○) : at 160 °C; (▲) : at 150 °C

The consideration of kinetic and thermodynamic parameters in the process of increase of length of effective conjugation allows the conclusion that the kinetics in both polymers is first order in nitrile groups. The values of effective activation energy of the process for PAN-c and PAN-r fall close together and equal 86.7 and 77.9 kJ/mol, respectively (20.7 and 18.6 kcal/mol). At the same time, the growth rate constant for PAN-c is considerably higher than for PAN-r. For instance, at 180 °C these constants are 14.5×10^{-5} and 4.3×10^{-5} s^{-1}, respectively. This fact is due to the difference in the preexponential, viz. 2.20×10^6 and 2.14×10^4 s^{-1} for PAN-c and PAN-r, respectively. Bearing in mind the close values of density for both polymers, the difference in preexponentials should be regarded as a result of variation in the steric factor. This is reflected by the activation entropy of the process of changes of the system of conjugated bonds. Thus, for PAN-c $\Delta S^{\ddagger} = 135.44$ kJ mol^{-1} K^{-1} (32.35 kcal mol^{-1} K^{-1}), whereas for PAN-r $\Delta S^{\ddagger} = 173.58$ kJ mol^{-1} K^{-1} (-41.46 kcal mol^{-1} K^{-1}). Since entropies of final state, i.e., of rigid conjugation sections, of both polymers are to be assumed to fall close together, the difference in $\Delta S^{\ddagger}$ values should result from the difference in the entropies in the initial state of the polymers. This is corroborated by the fact that, according to X-ray structure analysis data, PAN-c has a higher crystallinity than PAN-r. As should have been expected for the processes of PAN-c and PAN-r transformation, $\Delta H^{\neq}$ values do not differ greatly and equal 85 and 73.3 kJ/mol respectively (20.3 and 17.5 kcal/mol).

We have considered the effect of chain configuration of the polymer on the process of PCS formation. The macromolecular conformation also plays an important role in the process. Indeed, thermal transformation in oriented PAN films leads both to the acceleration of the process of PCS formation and to an increased efficiency of conjugation. In this respect the effect of IR irradiation on PAN is worthy of note[242, 243]. In general terms, the process of formation of conjugated C=N bonds caused by IR irradiation proceeds in a similar way to the thermal effect. The formation of C=N–H groups, resulting from the migration of the hydrogen from CH groups to C≡N groups, serves as the initiating step in this case as well. However, exposure to IR radiation is not accompanied by the formation of C=C bonds[242]. The accumulation of double bonds and the increase of length of effective conjugation results in an IR absorption at higher wavelengths, and this, in its turn, drastically increases the reaction rate. Thus, the process of PAN transformation under the effect of IR radiation proceeds with considerable self-acceleration. The irradiation of uniaxially oriented PAN films gives a polymer with a distinct anisotropy of optical properties, dichroism in the visible spectral region in particular. Figure 8 presents dichroism curves $[D = f(\lambda)]$ at various angles (φ) between the polarization plane and the orientation axis. The same figure shows the dependence $D = f(\varphi)$ at λ = 400 nm for a uniaxially oriented film.

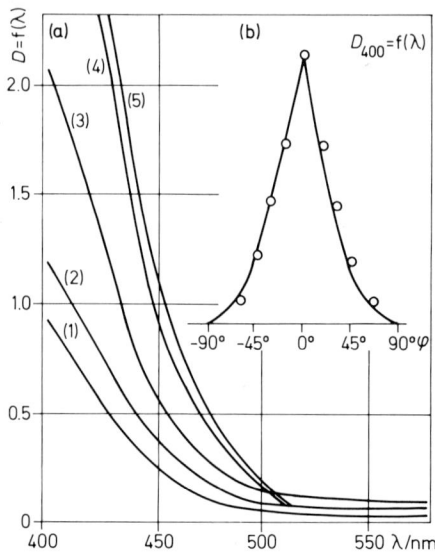

Fig. 8. Dichroism of electronic absorption spectra of oriented and nonoriented PAN films after IR irradiation. (a) Dependence of optical density (D) on the wavelength for various values of φ (angle between film orientation axis and light polarization plane). (1)–(3) : oriented PAN films; (1) : φ = 90°; (2) : φ = 45° (3) : φ = 0°; (4)–(5) : nonoriented PAN films; (4) : φ = 90°; (5) : φ = 0°. (b) Dependence of optical density $D_{400\ nm}$ on φ for oriented PAN films

The dichroism coefficient, calculated according to the formula $\rho = (D_\parallel - D_\perp)/(D_\parallel + D_\perp)$ (where $D_\parallel$ and $D_\perp$ stand for optical density when φ = 0° and φ = 90°, respectively), may reach a value of 0.5–0.6 at high orientation degrees. The fact that the absorption of plane-polarized light is maximal only if the polarization plane coincides with the direction of orientation of the macromolecules, as well as the fact that orientation favors the formation of longer effective conjugation blocks, proves that the formation of a PCS can be directed along the main chain by changing the conformation of the macromolecules. This, in particular, reflects the matrix character of the process under consideration.

3. Effect of Various Factors on the Conjugation Efficiency in PCSs

3.1. The Length of Conjugated Blocks

It was mentioned earlier that the conjugation efficiency in PCSs is determined by the degree of overlap of wave functions of the π-electrons and, therefore, should depend on the conformation of the polyene chain. It may be thought that the configuration of separate fragments should manifest itself to a degree which may affect the conformation of conjugation blocks. Effective conjugation always includes, in PCSs, considerably smaller sequences of units than one could have expected from the values of the molecular weight of a polymer[40, 41, 246, 247]. The reasons for that are as follows. Considering a molecule as an isolated system, the origination and elongation of the effective conjugation block at the initial step is energetically advantageous, since it is accompanied by a decrease in the internal energy of the system due to the gain in conjugation energy. However, the value of this "gain" decreases with the addition of subsequent units to the chain and for a length of the block of about 10–12 monomeric units it does not exceed 4.2–8.4 kJ/mol (1–2 kcal/mol). On the other hand, the process of elongation of a rigid conjugation system is not advantageous as far as the change in the entropy factor is concerned, since the number of possible conformations decreases in this process far more strongly than in the case of the formation of a polymer affording a greater number of conformers. Therefore, having reached a certain length of a coplanar conjugation block (not exceeding 10–12 units), the coplanar addition of the following units becomes energetically less advantageous than the formation of a new block of PCS which is not conjugated with the preceding one.

Calculations have shown that for the conservation of the coplanar arrangement, the sequence in a polyconjugated system cannot contain more than 30–33 carbon atoms. With longer sequences, thermal oscillations of the frame are sufficient for the withdrawal of the system from coplanar arrangement. The length of conjugated blocks in real systems is, as a rule, lower. Steric hindrance of the coplanar arrangement of the polyene chain arising from an interaction of side substituents should also be taken into account. Finally, intermolecular interactions in the solid phase and solvation effects in solutions should also, as a rule, serve as a factor which does not favor the conservation of the coplanarity of long fragments of the chain, whereas in some instances intermolecular effects may increase the conjugation efficiency. The above concepts make it possible to consider the polymeric chain on the whole as a system consisting of mutually noncoplanar blocks, each of them comprising coplanar units, or as a sort of a helical structure in which the planes of C=C-double bonds of each unit are shifted with respect to each other. Since the polymers under discussion, having identical chemical structures and equal molecular masses, may differ in length and in the character of the distribution of conjugation blocks, and since these differences are manifested in a complex of chemical and physicochemical properties, we can speak of an isomerism in PCSs of a type peculiar to this kind of polymers[41].

3.2. Effects of Conformational Transformations

The influence of conformational transformations on the conjugation efficiency[248, 249] may be exemplified by the data given in Figs. 9 and 10. Fig. 10 shows to the polarization degree, P, of poly(propynoic acid) vs. the pH of the solution. The data show that the change of pH from 3.5 to 12.5, affecting interaction of carboxylate anions (due to the dissociation of the carboxygroups), leads to the variation of the degree of straightening of the chain of poly(propynoic acid), the straightening reaching a

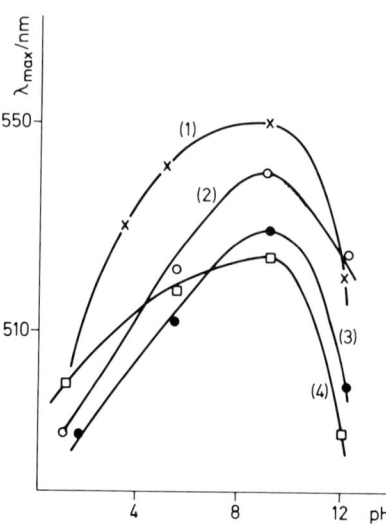

Fig. 9. Variation of λ_{max} in the fluoroscence spectra of solutions of poly(propynoic acid)s in water [(1) and (2)] and methanol [(3) and (4)] as a function of the ph of the solution. (1) and (3): Poly(propynoic acid) obtained by polymerization in liquid phase (PPAL); (2) and (4): poly(propynoic acid) obtained by polymerization in solid phase (PPAS)

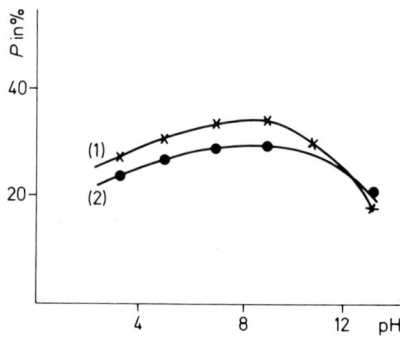

Fig. 10. Polarization degree (P) as a function of the pH of the medium in aqueous solutions. (1): PPAS; (2): PPAL

maximum at pH 9. Figure 9 demonstrates that the conjugation efficiency is highest just at a pH corresponding to the maximum straightening of the polyene chain of poly(propynoic acid). An increase of the ionic strength of the solution (*e.g.*, NaCl solution) suppresses to a certain degree the electrostatic interaction of the carboxylate anions, thus decreasing the straightening degree of the polyene chain and causing a less distinct effect of the pH of the medium on the conjugation efficiency.

The transfer of PCSs from solutions into the solid state may be accompanied by the origination of hydrogen and "salt" bonds, by associations in crystalline regions, or by charge transfer states and some other phenomena. These effects are followed by some conformational transformations in the macromolecules. The solution of the problem of the influence of these phenomena on the conjugation efficiency and on the complex of properties of the polymer is of fundamental importance.

We have shown with poly(barium propynoate)s as an example [40, 41] that the conjugation efficiency may vary considerably, depending on the presence of inter- or intramolecular "salt" bonds in the polymers.

Figures 11 and 12 contain data testifying to the change in conjugation efficiency following the formation of a solid phase by polymers of propynoic acid. It follows

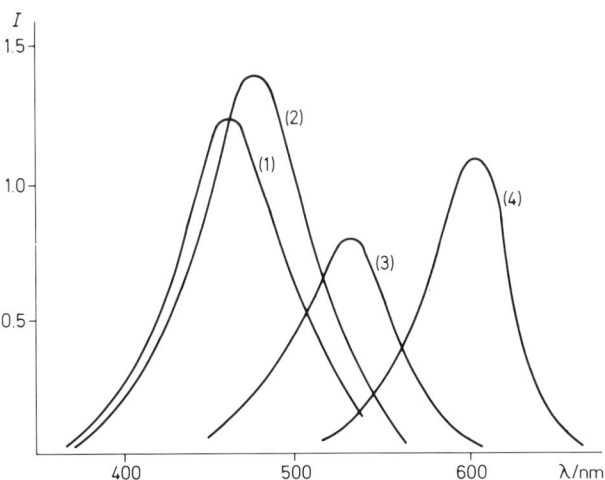

Fig. 11. Change of the fluorescence spectra of poly(propynoic acid)s as a result of transfer from solution [(1) : PPAL, (2) : PPAS] to the solid phase [(3) : PPAL, (4) : PPAS]. Concentration of aqueous solutions 10^{-4} mol/l. The fluorescence spectra of the samples in solid phase were taken from powdered polymers placed between two quartz plates. Slit width: 0.2 mm; sensitivity: 3,0; $\lambda_{excitation}$ = 365 nm. I = intensity

from these data that the polymer of *trans-transoid* configuration gives rise to a greater increase in conjugation efficiency than the *trans-cisoid* conformer in the process of solid phase formation. When compared to the changes in the IR spectra, these results give reason to believe that this effect is connected with conformational transformations resulting from the development of intramolecular hydrogen bonds in the first case and intermolecular hydrogen bonds in the second. Similar phenomena were

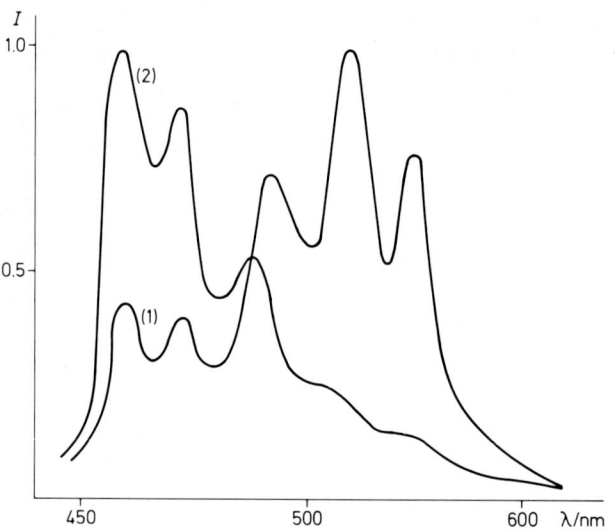

Fig. 12. Fluorescence spectra of DPAcN in solution (1) and in solid phase (2). Spectra of solid-phase DPAcN (5 % double bonds) were taken from the powder placed between two quartz plates. $\lambda_{excitation}$ = 324 nm

noted for the formation of salt bonds[40, 41]. Along with this, polymers containing heteroatoms in the main chain or in side substituents often reveal the reverse, i.e., the development of a solid phase leads to a higher conjugation efficiency. This effect seems to arise from the presence of the heteroatoms and from the possibility of generation of charge transfer complexes in the solid state. We have expressed an opinion that the development of crystalline zones in polymers may lead to a decrease in conjugation efficiency since crystallization taking place in accordance with known laws should not necessarily require conservation of coplanarity of conjugation blocks[250]. On the contrary, when steric factors, determined by the presence and the size of the substituents, are the reason for requiring conformation transformations of the main chain for a compact packing in the crystalline domains of the polymer, such transformations may occur even if they involve some perturbation of coplanarity of the conjugation blocks. The perturbation of coplanarity may also arise from the introduction of separate sections of conjugation blocks into various chaotically located crystalline zones. Analysis of properties of a large number of polymers with conjugation systems belonging mainly to polyazines and poly(schiff base)s[250] shows convincingly that properties typical of PCSs such as paramagnetism, absorption bands in electronic spectra, etc., are expressed to a substantially greater degree in amorphous PCSs than in crystalline ones. As regards the chemical properties, these differences are manifested, for instance, by a greater tendency of the amorphous polymers to form charge transfer complexes with electron acceptors. In order to find out directly the influence of crystallinity on the conjugation efficiency, polycondensation of hydrazine with various pairs of dialdehydes and diketones has been carried out. Each of these dialdehydes and diketones individually produces a crystalline polyazine which, due to this fact, has a low conjugation efficiency. It turned out that these polymers are deeply coloured, they contain $10^{17}-10^{18}$ paramagnetic centers per gram, they also form electroconductive charge transfer complexes, and show a high background level in their IR spectra. That means that amorphous copolymers of this type have properties typical of PCSs with a high conjugation efficiency. Thus, in this direction one of the possibilities of a systematic change of conjugation efficiency of PCSs is opening.

3.3. Effect of Separation of Conjugated Blocks

Studying the effect of different factors on conjugation efficiency in PCSs, we have disclosed an interesting phenomenon called the "effect of separation of conjugated blocks"[251, 252]. The effect consists in the fact that conjugated blocks separated by saturated sections (in graft and block copolymers) or, in a number of cases, PCSs with inert fillers, exhibit properties typical of PCSs to a greater degree than polymers containing only polyconjugated blocks[251, 252]. The most detailed studies of this effect were carried out on block copolymers of butadiene and phenylacetylene and on graft copolymers of acrylonitrile and ethylene-propylene copolymer (EPC). These copolymers have been subjected to thermal treatment. Thus, under certain conditions the thermal transformation of PAN (180° C, 6 h) results in the formation of products with λ_{max} reaching from to 480 to 490 nm with a concentration of paramagnetic centers not exceeding $5 \times 10^{17} - 7 \times 10^{17}$. At the same time a similar transformation of a graft copolymer of acrylonitrile and EPC leads to products with λ_{max} in the region of 530 to 540 nm with a concentration of paramagnetic centers amounting to $3 \times 10^{19} - 4 \times 10^{19}$, i.e., almost two orders of magnitude higher than in the case of poly(acrylonitrile), irrespective of the fact that the concentration of PAN blocks responsible for paramagnetism does not surpass 20–40% in these copolymers. Moreover, after separation of thermally transformed PAN blocks attached to EPC through ester groups (separation, e.g., by saponification), the conjugation efficiency of PAN blocks drops sharply and does not differ from that of products resulting from a respective thermal transformation of poly(acrylonitrile). The detachment of the conjugated blocks is accompanied by changes in the thermomechanical properties of the copolymer. Thus, in particular, the region of high elastic state of polymers narrows (due to the decrease in flow temperature with glass transition temperature remaining constant) and becomes similar to that of EPC. The reason for the existence of the effect of separation of conjugated blocks in polymers with conjugation systems cannot be claimed as being fully understood at present. We believe, however, that this effect results from the improved conditions for the conservation of coplanarity of the conjugated blocks in case of separation. In other words, we proceed from the assumption that intermolecular interactions of PCSs bring about mutual "distortions" of the conjugated blocks, hampering the manifestation of the highest possible conjugation efficiency. This means that conjugated blocks in isolation are more capable of effective conjugation than under conditions of intermolecular interactions. From this standpoint, the decrease in conjugation efficiency arising from crystallinity, described above, could be considered as a particular case of a general trend.

The utilization of the effect of separation of conjugation blocks represents yet another possibility for the control of conjugation efficiency in PCSs.

One cannot rule out, however, the possible assumption that this effect is a consequence of an interaction of conjugated blocks in graft and block copolymers that may lead to the overlap of wave functions of various conjugated blocks, followed by an increase in conjugation efficiency in the polymers.

3.4. Electronic Excitations and Energy Transfer in PCSs

PCSs are systems of chromophores bound into a single macromolecule. Therefore, the study of processes of electronic excitation and energy transfer, as well as the investigation of the ways of deactivation of excited states, should lay a foundation for the understanding of such properties of PCSs as reactivity in photochemical transformations, photosensitizing and photoelectric activity, photoinitiated paramagnetism, etc.

We have found for polypropynoic acid that this series of polymers reveals selective fluorescence spectra together with nonselective absorption. To account for this phenomenon, a scheme was proposed according to which PCSs are characterized by energy transfer from excited levels of some conjugation sections to the lower levels of other sections, followed by luminescence from the latter[40, 41, 246, 248, 249, 253].

Similar results were obtained by Benderskii[254] for poly(phenylacetylene). His data show that the position of the maximum in the fluorescence spectra reflects the presence of conjugated sections with the lowest energy of the excited states, *i.e.*, sections with the highest conjugation efficiency. This gives a practicable possibility for evaluation of the relative conjugation efficiency of a system *via* determination of the character of the shift of luminescence spectra, accompanying the introduction of a series of luminescent additives to the solution[253].

It is to be noted that, in contrast to the fluorescence spectra of solutions of dehydrogenated poly(acenaphthylene) (DPAcN) which do not depend on the wavelength of the exciting light (in the range from 304 to 436 nm), PCSs, such as poly-(propynoic acid), polyconjugated polyamines, or products of PAN thermal transformation, reveal a substantial shift of the maximum in the fluorescence spectra towards longer wavelengths with increasing wavelength of the exciting light thus manifesting the so-called bathochromic luminescence. The origin of this phenomenon is not clear yet, and requires further studies.

Fluorescence and phosphorescence spectra of poly(propynoic acid) (PPA), polyphenylene (PP), and DPAcN show that the difference of energies between the lower excited singlet and triplet states, as observed in the case of PP (583 nm) and DPAcN (528 nm), is considerably greater than that of poly(propynoic acid) (270–300 nm) which besides $\pi \to \pi^*$ transitions may undergo $\pi \to n^*$ transitions. PCSs showing only $\pi \to n^*$ transitions are characterized by a longer lifetime (τ_t) of the triplet states than that of PCSs which together with $\pi \to \pi^*$ transitions may undergo $n \to \pi^*$ transitions. These values equal 2.0, 1.05, and 0.5, respectively, for PP, DPAcN, and PPA. It is possible that the contribution of excited states caused by $n \to \pi^*$ transitions accounts basically for a bathochromic luminescence of some PCSs and for a shift of the maxima in the luminescence spectra of polymers of this kind when proceeding from the solution to the solid phase. PCS solutions reveal concentration-quenching accompanied by a decrease in lifetime of the excited states. Thus, poly(propynoic acid) displays a decrease in lifetime of excited states from 4×10^{-8} to 2×10^{-9} s with the concentration of the polymer in solution growing from 1×10^{-4} to 1×10^{-2} mol/l. A concomitant bathochromic shift is probably due to the fact that during the lifetime the excited state changes its carrier several times, each transfer act having a certain probability of nonradiative deactivation.

As the concentration of solutions increases, fluorescence depolarization occurs, in addition to self-quenching. Since concentration quenching is almost the same both in the solutions of various viscosity and in solid matrices[253], one should arrive at the conclusion that this process is not controlled by diffusion. In as much as concentration quenching, accompanied by a bathochromic shift in the fluorescence spectra, is not characterized by any changes in the absorption spectra, that may mean that this process involves the formation of excimers. The overlap of absorption and emission spectra typical of PCSs implies that a certain contribution to the process under discussion is also due to a dipole-dipole mechanism of the energy transfer. It is also to be noted that PCSs have a threshold concentration (10^{-5} mol/l) after which a steady shift of the emission band begins. An analysis of the emission spectra of DPAcN of various dehydration degrees, and of acenaphthylene and phenylacetylene copolymers of different compositions (Fig. 13) shows that the processes of intramolecular energy transfer are determined to a considerable degree by the nature and length of the saturated sections separating the conjugated blocks. If these sections are sufficiently long, the conjugated blocks will act as separate chromophores. As a result of this, the fluorescence spectra will contain separate bands of emission of interacting conjugated blocks.

A distinctive feature of this type of polymer is that the processes of intermolecular transfer of excitation energy are also hindered to a considerable degree since they become noticeable only at very high (of the order of 10^{-3} mol/l) concentrations of

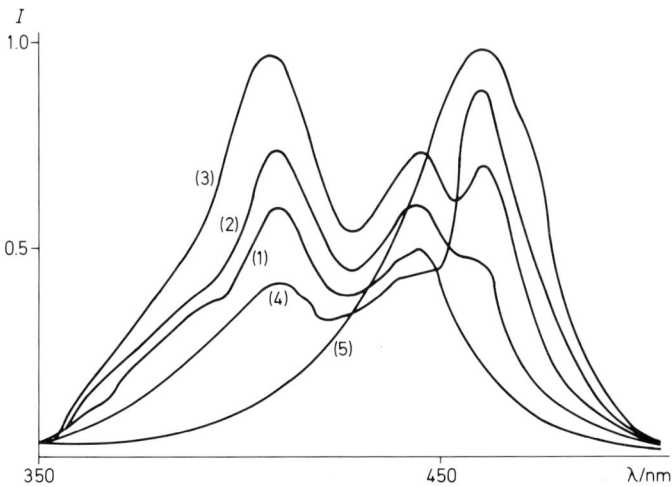

Fig. 13. Fluorescence spectra of copolymers of acenaphtylene with phenylacetylene. Content of phenylacetylene blocks: (1) : 14 mol-%, (2) : 20 mole-%, (3) : 35 mole-%, (4) : 48 mole-% (5) : 94 mole-%. Spectra are taken in benzene solution $C = 10^{-4}$ mol/l. $\lambda_{excitation} = 324$ m

polymers. Along with this, the introduction of an additive not bound to polymer chains into the solution of a polymer of this kind (*e.g.*, phenylacetylene homopolymer) is accompanied by a highly effective energy transfer to this additive even though the concentration of the latter is about two orders of magnitude lower than that of the conjugated blocks separated by saturated sections. Apparently, the processes involving energy transfer also reveal the effect of separation of conjugated blocks found and noted by us in the course of investigation of other phenomena peculiar to PCSs[251, 252].

In order to clear up the mechanism of inactivation of excited states, we examined the processes of quenching of fluorescence and phosphorescence in PCSs by the additives of the donor and acceptor type[253, 255, 256]. Within the concentration range of $1 \times 10^{-4} - 1 \times 10^{-3}$ mol/l, a linear relationship between the efficiency of fluorescence quenching $[(I_0/I) - 1]$ and the quencher concentration was found. For the determination of quenching constants, the *Stern-Volmer* equation was used, viz. $I_0/I = \tau - K_g I_0 [Q]$, in which K_g is the bimolecular rate constant of interaction of quencher Q with the excited states of the PCS, τ is the lifetime of excited molecules with no quencher, I_0 is the quantum yield of fluorescence in the absence of the quencher, and I is the quantum yield of fluorescence in the presence of the quencher.

The data in Figure 14 imply that the efficiency of fluorescence quenching by acceptors grows simultaneously with an increase in electron affinity of the quenchers.

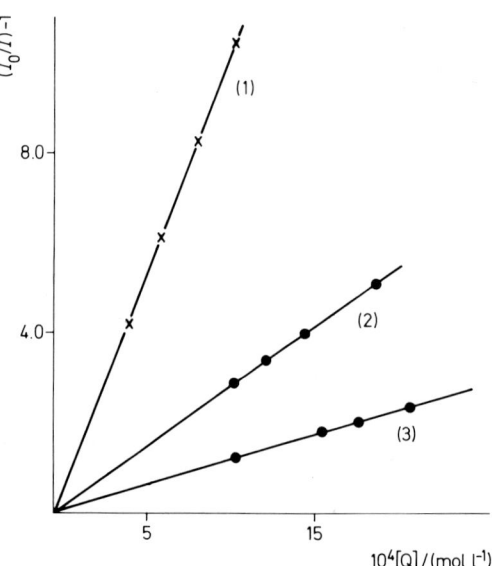

Fig. 14. Quenching of fluorescence of PP solution in benzene. Quenchers (Q) : (1) : tetracyanoethylene, (2) : chloranil, (3) : trinitrobenzene I : fluorescence in relative units in the absence of a quencher, I : in the presence of a quencher

This makes us conclude that the process of quenching is associated with an electron transfer. The efficiency of phosphorescence quenching by acceptors follows, as well, the growth of electron affinity of the latter. Phosphorescence quenching constants are two orders of magnitude lower than fluorescence quenching constants. This indicates a higher efficiency of electron acceptors with respect to singlet states of PCSs than with respect to triplet ones. The comparison of luminescence and absorption spectra implies that the process of quenching is of both a static and a dynamic character. The latter circumstance is also evidenced by a shorter life time to singlet and triplet states in the presence of acceptors.

The studies of fluorescence quenching processes in PCSs caused by electron donors (N,N-dimethylaniline, triethylamine, and aminostyrene) show that quenching constants of singlet and triplet states grow with an increase in ionization potential of the donors. Generally speaking, electron donors are substantially less effective quenchers of excited states than electron acceptors. One of the reasons for this is the fact that donor quenching excludes a dipole-dipole mechanism of excitation transfer, since the absorption spectra of PCSs and electron donors studied by us do not overlap. The fact that PCSs manifest donor properties to a greater degree than acceptor properties may serve as another reason, thereby reducing the efficiency of static deactivation of excited states resulting from the formation of associations.

It is to be stressed that the action of quenchers is also accompanied by the effect of separation of conjugated blocks mentioned above. Thus, in the case of trinitrobenzene as a quencher, a concentration of trinitrobenzene of 1×10^{-3} mol/l is quite sufficient for complete quenching of DPAcN fluorescence (7% double bonds, $\tau_0 = 2.5 \times 10^{-8}$ s), but brings about only an insignificant degree of quenching of the PP solution ($\tau_0 = 9.0 \times 10^{-3}$ s), the concentration of chromophoric groups being the same[253, 255, 257].

We have payed much attention to the investigations of excitation energy transfer in PCSs since the analysis of disclosed effects affords an entrance to the study of certain photochemical processes in PCSs. As to the photochemistry of PCSs, it is considered one of the promising fields of study not only from the theoretical but also from the practical point of view.

4. Chemical Properties of PCSs

4.1. Reactions with Electrophiles and Nucleophiles

The reactions of PCSs with electrophilic agents were studied by us, e.g. the oxidation and chlorination of polyacetylene[39–41]. Oxygen attack leads to the degradation of the PCS in the polymer. The process is accompanied by the formation of carbonyl and ester groups and by the amorphization of the polymer. The chlorination of polyacetylene at room temperature yields a white polymer containing up to 65% chlorine. Together with the chlorination, crosslinking of the polymer occurs. A technique for determination of the molecular mass of the polymer based on the examination of its chlorination products has been proposed[39].

The interaction of PCSs with nucleophilic reagents was studied by us; we took the reactions of hydrazine and phenylhydrazine with polyazines, poly(schiff base)s, and other polymers containing conjugated C=N bonds as an example[40, 41, 117, 258].

The addition of hydrazine (and phenylhydrazine) to these polymers proceeds with the opening of the C=N bonds. If the main chain consists of appropriate fragments, this reaction is followed by the degradation of this chain.

Destructive hydrazinolysis is of importance for the establishment of the structure of the PCS main chain, since the reaction products can easily be identified. The importance of this reaction also results from the fact that C=C bonds do not undergo hydrazinolysis. A general scheme of the process of destructive hydrazinolysis may be given as follows:

$$\left[=N-N=C-R'-C= \atop RR \right] + n\ H_2N-NH_2 \longrightarrow n\ H_2N-N=C-R'-C=N-NH_2 \atop RR$$

The degree n of completion of the reaction is determined to a considerable degree by the structure of the polymer (crystallinity, solubility, etc.) and by the conditions of the process. Thus, the introduction of 10–15% of dimethylformamide into the solution allows the process to be conducted with a quantitative yield. Except the above mentioned polymers, polynitriles, products of PAN thermal transformation, paracyanogen, polyquinolines, and polypyridines also can undergo destructive hydrazinolysis, thus making it possible to use destructive hydrazinolysis as a general method of determining the structure of nitrogen-containing PCSs in the same manner as ozonolysis serves as a general method for the analysis of polymers containing C=C bonds in the main chain.

4.2. Thermal Stability. Pyrolysis Reactions

We have noted that the thermal stability of PCSs arises from the low enthalpy of these polymers, resulting from the gain in conjugation energy in the process of their formation. In the following we shall discuss the features of PCS degradation processes. The most general peculiarity of these processes is that they are accompanied by self-stabilization of the polymer, as follows from the stepwise shape of the kinetic curves of weight losses. This effect is due to the "perfection" of the conjugation system and to the formation of crosslinked structures. In contrast to some other classes of PCS, the products of thermal degradation of poly(schiff base)s are characterized by elevated nitrogen content[40, 41, 126, 127]. The replacement of aliphatic fragments in the main chains of these polymers by aromatic groups increases the general heat resistance. The same effect is observed in the case of substitution of phenylene fragments by pyridyl groups. It is to be noted that when poly(schiff base)s are characterized by a certain crystallinity, the amorphization occurs only at deep degradation steps.

Degradation of polyazines proceeds in quite a different way[40, 41, 117]. The presence of NN-bonds in the polymers predetermines the fact that their thermal degradation is always accompanied by the evolution of nitrogen. According to the peculiarities of this process, polyazines can be divided into three groups. Polyazines with a continuous system of conjugated bonds in the main chain should be assigned to the first group. Degradation of polymers of this type is characterized by a rapid weight loss in a relatively narrow temperature range (20–30 °C), the loss in some cases reaching 50%, followed by a gradual and relatively moderate loss up to 550–570 °C. Thermal degradation of polyazines of this type occurs with marked exothermal effects, accompanied by a practically complete evolution of nitrogen as N_2 from the polymer.

Polyazines with conjugation sections separated by heteroatoms (O, S, N) behave in a substantially different way in the process of heating[259]. Thermal degradation of polyazines of this type occurs gradually and begins at relatively high temperatures (300–500 °C). Thus, polyazines offer an example of a case in which an interruption of conjugation is not accompanied by the decrease of thermal stability; on the contrary, this leads to a marked growth of stability.

Bearing in mind the occurrence of a period of rapid pyrolysis of polyazines with a continuously conjugated chain, and the fact that gaseous products of intense degradation consist to 65–70% of elementary nitrogen, it should be concluded that the thermal degradation of these polymers occurs in a similar way to the degradation of azo compounds. This allows the assumption that these systems are characterized by an azo-azine mesomerism by which the single bond character of the CN-bond is increased by the participation of the azo form as mesomeric structure that determines the nonuniform kinetics of their pyrolysis.

This mesomerism can be represented as follows:

$$\left[=N-N=C-C= \atop RR \right] \longleftrightarrow \left[-N=N-C=C- \atop RR \right]_n$$

The interruption of conjugation by heteroatoms[40, 41, 259] leads to a higher heat stability for the particular reason that in these cases the possibility of azo-azine rearrangement is excluded.

Thermal stability of polyazines of the third group, i.e., polymers with conjugated sections separated by aliphatic blocks, is determined by the thermal stability of the latter. The degradation process takes place gradually, and it is not accompanied by any considerable heat effects[41, 117].

Thus, the presence of a continuously conjugated system in polyazines, a system including several fragments of structure, does not increase the thermal stability of the main chain irrespective of the decrease of the internal energy of the system. On the contrary, the availability of conjugation and the resulting capability of the system to undergo azo-azine rearrangement are responsible for the occurrence of a deep degradation of polymers of this kind at relatively low temperatures.

Pyrolysis of thermally transformed products of poly(acrylonitrile) involves, basically, open chain sections, the conjugation blocks exerting a catalytic influence on the process. This catalytic influence has been proved to be mainly of intramolecular character[232]. Therefore, deeper cyclization leads, on the one hand, to an increased general heat stability of the polymer (as evidenced by the decrease in amount of volatile matter) and, on the other hand, to a decreased effective activation energy, the value of activation energy falling with increasing content of conjugation blocks in the polymer.

We have studied the effect of the stereoregularity of the initial PAN on the pyrolysis of products formed by cyclization of the polymer[238, 239, 241]. A comparison of the induction period of the process of degradation of noncyclized PAN fragments with a certain conjugation efficiency in the polyconjugation blocks shows that the induction period declines as the conjugation efficiency grows, irrespective of the way the cyclization was effected and of the degree of stereoregularity of the initial polymer. In particular, when λ in the luminescence spectra reaches 600–610 nm, the degradation of PAN cyclization products (no matter how cyclization was carried out) occurs with no induction period.

Thus, it should be suggested that there exists a certain, quite definite, value for the critical length of effective conjugation catalyzing the thermal degradation of PAN regardless of its microtacticity or the previous history of the sample. The microtacticity of the initial PAN manifests itself only to a degree which is determined, as was mentioned above, by the effect of the microtacticity on the kinetics of the cyclization of nitrile groups. The thermal stability of poly(phenylacetylene) with respect to its molecular mass and the degree of branching was analyzed in Ref.[260].

We have studied the influence of conjugation in the main chain of a PCSs on the reactivity of functional side groups in the case of poly(propynoic acid) (PPA) and its copolymers[261–263]. The study of the kinetics of PPA decarboxylation has demonstrated that this process takes place at lower temperatures (145 °C) than in the case of its hydrogenated analogue, poly(acrylic acid) (200 °C). E_{eff} of PPA decarboxylation is also considerably lower than that of poly(acrylic acid) decarboxylation [21–42 kJ/mol and 184 kJ/mol, respectively (5–10 and 44 kcal/mol)]. It is noteworthy that PPA of *trans-transoid* configuration undergoes decarboxylation with E_{eff} = 21 kJ/mol (5 kcal/mol), whereas for the polymer with prevalent *cis-transoid* or *trans-cisoid* configuration this value is twice as high [38.5 kJ/mol (9.2 kcal/mol)]. These results conform to the fact that in the solid state PPA of *trans-transoid* configuration is characterized by a higher conjugation efficiency as is evidenced by luminescence spectra.

Together with the decarboxylation, dehydration of PPA takes place on heating. Examination of the decarboxylation of copolymers of acrylic and propynoic acids having blocks in their structure has revealed two interesting phenomena[261, 262]. First, these copolymers undergo decarboxylation more readily than any of the homopolymers. Second, decarboxylation involves the units of acrylic acid at temperatures which do not affect homopolymers of acrylic acid. In our view, the first phenomenon is accounted for by the effect of separation of conjugation blocks exemplified by this particular chemical reaction. As to the second observation, we believe that decarboxylation under relatively mild conditions (160–170 °C) affects, apparently, the fragments of acrylic acid located at the junctions of the blocks.

Decarboxylation of copolymers of propynoic acid and phenylacetylene of varying composition was found to manifest an increase in reactivity of the carboxylic groups with increasing concentration of conjugated double bonds in the chain.

4.3. Polyconjugated Polyelectrolytes

In the following we shall consider the effect of polyconjugation and configuration of the polyene chain on the equilibrium dissociation of polyconjugated polyelectrolytes[40, 41, 45, 264], such as PPA. The comparison of pK_a and Z(energy of proton dissociation) for PPA having various configurations of the polyene chain, shows that the polymer with *trans-transoid* configuration is a weaker acid than the polymer with *trans-cisoid* or *cis-transoid* configuration. The pK_a values for these polymers are 6.4 and 5.3 and the Z values are 33.7×10^{-14} and 21.6×10^{-14} erg respectively (1 erg = 10^{-7} J). The values of constants characterizing dissociation equilibria in the Kern, Katchalsky, and Karuba equations and the concentration dependence of these constants are given in our works[40, 41, 45]. Since spectral data imply that these polymers in solutions do not practically differ in conjugation efficiency[246], the differences pointed out should be considered as a result of the configuration of the polyene chain having carboxylic groups on one side in the case of a *trans-transoid* configuration and on two sides in the other polymers. Indeed, the electrostatic field resulting from the dissociation of the carboxylic groups in the *trans-transoid* polymer (carboxylic groups on one side of the chain) should hamper to the greatest degree the dissociation of the following groups. This is an example of the situation when conjugation in the main chain manifests itself not as such but through the rigidity of the main chain and, consequently, the existence of stable *cisoid* and *transoid* conformations. In general, conjugation in the main chain is accompanied by a reduction of the acidity of the polyacid, as can be seen from the comparison of pK_a values of poly(propynoic acid) and poly(acrylic acid) (pK_a 4.8).

The salt effect is very strong in polyconjugated polyelectrolytes. Figure 15 is a graph of the proton dissociation energy vs. the dissociation degree of PPA of different structures. Also, the graphs for poly(methacrylic acid) and a copolymer

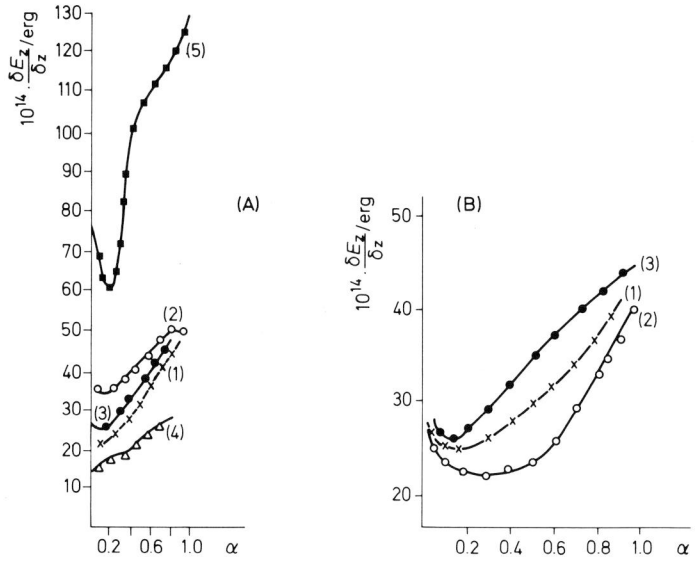

Fig. 15. Energy of proton dissociation (E_z) from Z times ionized polyelectrolyte molecules as function of the degree of dissociation (α). (A) – PPAL (1), PPAS (2), PPA (3), poly(methacrylic acid) (4), copolymer of acrylic acid with ethylenesulfonic acid (50 : 50) in aqueous solutions (5), (B) – PPAL (1), PPAS (2), PPA in the presence of NaCl (3) (●) : [NaCl] = 0; (x) : [NaCl] = 0.25 mmol/l; (○) : 0.50 mmol/l

of acrylic acid and ethylenesulfonic acid are shown. The data indicate that the dependences of poly(methacrylic acid) and PPA are substantially different. In the case of poly(methacrylic acid), the graph reflects the unfolding of the polymer coil, whereas the macromolecule of the polyconjugated polyelectrolyte in solution has a comparatively rigid linear shape, with a portion of carboxylic groups (in accordance with the pK_a value) already dissociated.

However, the addition of even small volumes of alkali leads to the screening of these groups, with a subsequent decrease of the proton dissociation energy at low dissociation degrees. This complies with the salt effect (Fig. 15).

4.4. Photoinitiated Reactions

We have considered the factors determining the participation of PPA in photochemical processes[40–42, 45]. Poly(propynoic acid) has served as an example of the perturbation of a conjugated system resulting from photoinitiated oxidation. At the initial stage of exposure to "white" light, the breakdown of the longer conjugation blocks occurs, followed by an accumulation of shorter ones (Fig. 16). On the other hand, oxidation under UV irradiation leads to the breakdown of long and short conjugation sections (Fig. 17).

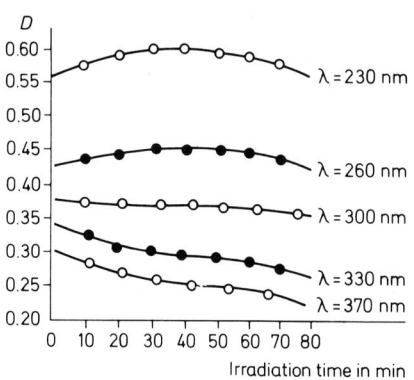

Fig. 16. Change of optical density (D) of aqueous solutions of poly (potassium propynoate) for different wavelengths in the process of oxidation during irradiation with "white" light

The reaction of photoinitiated reduction of polyphenylene with triethylamine [253] (wavelength of illuminating light 365 nm) has a quantum yield of 0.025 in methanol. The increased polarity of the solvent (azamethylformamide) results in a higher quantum yield. ESR spectra and pulse excitation experiments testify to the participation of ion-radicals in the reaction. In the absence of a reducing agent, polyphenylene undergoes photodegradation in nonpolar solvents. It was suggested that this process follows a two-photon mechanism. Polyphenylene photodegradation involves a triplet state, since the introduction of additives with the energy of the triplet state higher and lower than E_T of polyphenylene [167.5 kJ/mol (40 kcal/mol)] results in a higher or lower degradation rate, respectively[253, 266]. Oxygen [E_T = 96.3 kJ/mol (23 kcal/mol)] and chloranil [E_T = 293 kJ/mol (70 kcal/mol)] were taken as additives.

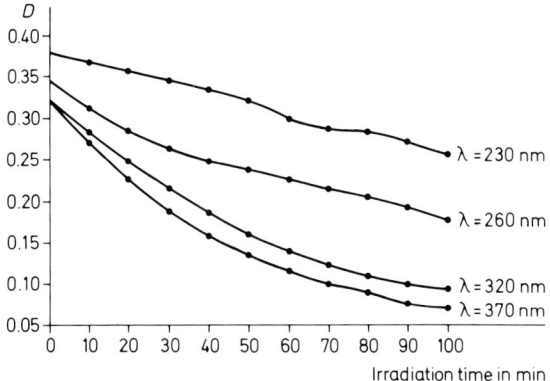

Fig. 17. Change of optical density of aqueous solutions of poly(potassium propynoate) for different wavelenghts in the process of oxidation during irradiation with UV light

4.5. Reactions with Dienophiles

We have also used poly(propynoic acid) in our studies of the photochemical interaction of PCSs with dienophiles, such as maleic anhydride, tetracyanoethylene, and styrene. This photochemical reaction of Diels-Alder type is accompanied by the breakdown of the conjugation system and the formation of slightly colored adducts[266]. Together with the cycloaddition reaction, photodegradation of PPA and its adducts takes place. A cycloaddition reaction is always preceded by the formation of a donor-acceptor complex of a PCS with a dienophile.

The comparison of rates of cycloaddition of maleic anhydride, tetracyanoethylene, and styrene to PPA shows that the latter, irrespective of the presence of electronegative groups, behaves in these reactions not as an "electron-poor" diene system. This fact, together with the composition of side products (giving evidence of PPA decarboxylation), allows the assumption to be made that the cycloaddition of dienophiles involves mainly decarboxylated polyene sections of *cis-transoid* structure[213, 266]. This is in agreement with the fact that PPA with predominant *trans-transoid* configuration interacts with these dienophiles at a substantially lower rate. The ultimate amounts of the dienophile combined with PPA of this structure is also considerably smaller.

4.6. Molecular Complexes and Their Electronic Spectra

Interaction of PCSs with electron acceptors and donors results in molecular complexes with partial or complete charge transfer. In particular, detailed investigations involved charge transfer complexes (CTC) of poly (schiff base), polyazines, products of thermal transformations of PAN and a number of other PCSs[129, 238, 241, 243, 267].

Electronic spectra of these complexes, as Fig. 18 shows, contain two new bands, one of which (long wave) is due to the acceptor absorption and the other

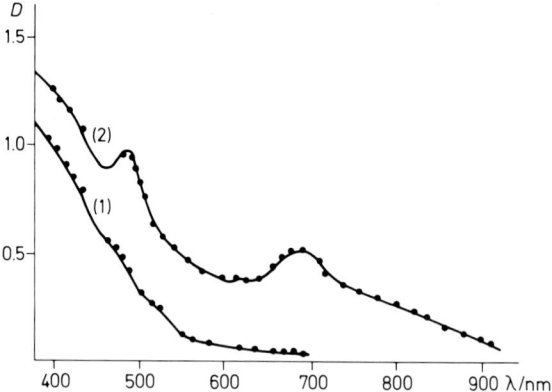
Fig. 18. Electronic absorption spectra of a poly(schiff base) obtained by polycondensation of 4,4'-diacetyldiphenyl sulfide and *p*-phenylenediamine (1) and of its complex with iodine (2)

results from the formation of states with charge transfer. The extent of charge transfer can be determined from the short-wave shift of the acceptor transfer band[273] or the location of the transfer band. Lower conjugation efficiency in PCSs results in a short-wavelength shift of the transfer band.

Since the energy of the transfer band is determined by the difference between the donor ionization potential and the acceptor electron affinity, this fact points to the increase of the PCS ionization potential with decreasing conjugation efficiency. Therefore, the location of the transfer band of the molecular complexes of an acceptor and various PCSs can serve as a criterion for the conjugation efficiency in the latter. In Refs.[267–272] the data for a number of molecular complexes are given, and the comparison with the electrical properties of the complexes is made.

The analysis of ESR and electronic spectra shows that the molecular complexes of PCSs with acceptors represent the states both with partial ($D^{\delta+} \ldots A^{\delta-}$) and complete charge transfer ($D^{+} \ldots A^{-}$). Depending on the chemical nature of the components and the conditions of the formation of molecular complexes, the ratio $[D^{\delta+} \ldots A^{\delta-}]/[D^{+} \ldots A^{-}]$ may vary within the range of 10 to 16^6. Since all molecular complexes of PCSs studied by us obey the Curie law, it should be believed that the complete charge transfer (which can be traced by ESR) occurs as early as in the ground state. This is a result of the high polarizability of the PCSs and a sufficiently great electron affinity of the acceptors used. On the other hand, we have demonstrated that the induced transition of molecular complexes from the state with a partial charge transfer to that with a complete charge transfer is possible[267]. Indeed, as is evidenced by Fig. 19, irradiation of a molecular complex in the transfer zone may result in an increase of concentration of paramagnetic centers, and hence of the concentration of complete transfer states, by a factor of 10^5. Figure 20 presents kinetic curves of growth and decrease of concentration of light-induced states with complete charge transfer. The time of concentration increase reaches from 25 to 55 min., depending on the complex.

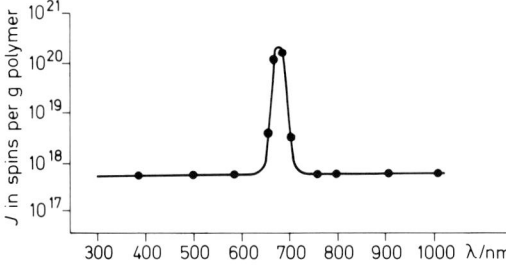

Fig. 19. Dependence of the intensity of the ESR signal on the wavelength of incident light of a complex of a poly(schiff base) (prepared from 4,4'-diacetyldiphenyl sulfide) with bromine; mole ratio 1 : 4.8

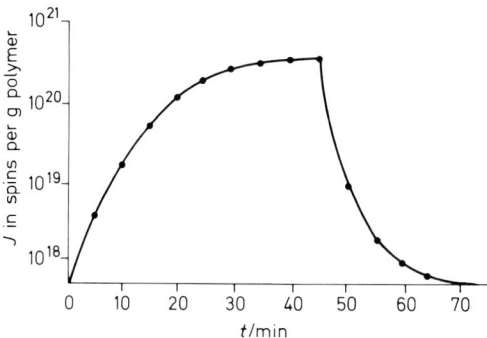

Fig. 20. Increase of the intensity of the ESR signal during the irradiation and its decay in darkness of a bromine complex of a poly(schiff base) prepared by polycondensation of 4,4'-diacetyldiphenyl sulfide and p-phenylenediamine; mole ratio 1 : 5.8

In these experiments, the entire concentration of spontaneously formed and photoinduced states, after the latter have reached ultimate values, becomes practically equal to the concentration of acceptor molecules.

Thus, irradiation in the transfer band results in practically the whole molecular complex being found in the complete transfer state. The course of the curve corresponding to the increasing ESR signal (Fig. 20) is typical of a number of PCSs and is well described by the relationship $\Delta I = I_{max}(1-e^{-t/a})$ where a is constant. The kinetics of decay of the concentration of paramagnetic centers differs substantially, depending on the PCS. Thus, poly(schiff base)s (Fig. 20) are successfully described by the equation $\Delta I = I_{max}/(1 + at)$ where a is a constant, but in the molecular complexes of products of thermal transformation of PAN the concentration of photoinduced paramagnetic conters remains for many months without changes. The structure of products of the thermal transformation of PAN seems to favor the emergence of local states (traps), the intensity of which is essentially greater than kT, thus giving rise to the stability of photoinduced states with complete transfer.

It should be noted that the properties of a CTC depend to a considerable degree on the conditions of their preparation. Temperature increase, in particular, favors the accumulation of complete charge transfer states in a CTC. In the case of a CTC obtained in solution, the increase of dielectric constant of the solvent has the same effect. The method of preparation of a CTC also affects the kinetic curves of the accumulation and depletion of complete transfer states arising at protoirradiation.

CTC formation is accompanied by a volume contraction[129]. Thus, poly(schiff base)s having no side groups give complexes with a density 10–15% higher than that calculated on an additivity assumption.

X-ray diffraction analysis of crystalline poly(schiff base)s and their low molecular models shows that the formation of molecular complexes is accompanied by an increase in interplanar distances and, in a number of cases, by complete amorphization. Molecular complexes of poly(schiff base)s with Br_2 decompose with time, because of the bromination of the donor components, forming C–Br bonds. Substitution of hydrogen by bromine in phenyl groups occurs only in cases in which these groups are not included into the main polymeric chain.

We have already pointed out that the reduction in conjugation efficiency in PCSs is followed by a short-wave shift of the CTC transfer band. This accounts for the fact that poly(schiff base)s and polyazines having conjugated sections separated by oxygen and sulfur atoms are characterized by a short-wave shift of the transfer band of CTC with all acceptors compared to the respective polymers having no interruption of the conjugated chain. This shift may reach 20–50 nm.

The products of the thermal transformation of PAN were used for the examination of the effect of stereoregularity of the initial polymer on the CTC properties. At equal transformation degrees, the concentration of complete charge transfer states in a bromine-complexed PCS, obtained on the basis of PAN-c (i. e., a polymer with elevated content of isotactic sequences), exceeds by two orders of magnitude this parameter in the polymer obtained on the basis of PAN-r. Finally, stereoregularity of the initial PAN also affects the disposition of a CTC obtained from this polymer to the formation of photoinduced states with complete charge transfer. Both the values of the stationary concentration of these states and the rate of growth to this level, are considerably higher for a PCS obtained from the polymer with elevated stereoregularity. All this characterizes the effect of PCS stereoregularity on their reactivity in the formation of a CTC. The semiconductive properties of PCS complexes of various classes with electron donors have been studied[267, 268].

5. Photosensitizing and Catalytic Properties of PCSs

5.1. Photosensitizing Activity

Low-molecular semiconductors and, in particular, phthalocyanines are known to exhibit photosensitizing properties[268]. The photosensitizing activity of these substances has been studied in most detail for the reaction of oxidation of ascorbic acid.

Bearing in mind the semiconductive properties of PCSs one might expect that these substances, being p-type semiconductors in air, possess photosensitizing activity. We, indeed, have demonstrated[40] that PCSs, such as poly(schiff base)s, salts of poly(propynoic acid), or polyquinoline, are active photosensitizers of

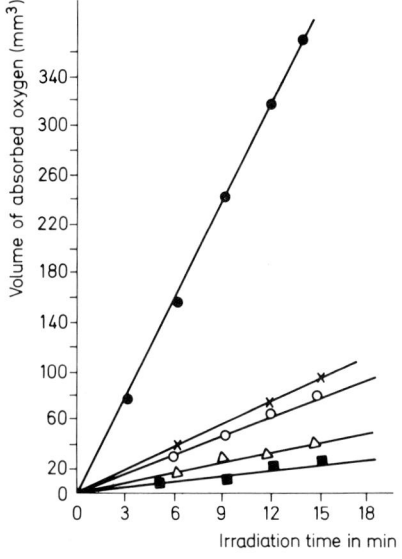

Fig. 21. Kinetic curves of photosensitized oxygen absorption by thermally treated PAN samples during irradiation with "white" light. (●) : PAN (450 °C, 3 h); (×) : PAN (350 °C, 3 h); (○) : PAN (200 °C, 30 h); (▲) : PAN (200 °C, 40 h); (■) : PAN (520 °C, 3 h)

ascorbic acid oxidation by oxygen. Figure 21 shows the kinetic curves[b] of photosensitized oxidations of ascorbic acid in the presence of a number of polymers[c].

According to Fig. 21, there is a clear-cut correlation (in any event for the products of the thermal transformation of PAN between semiconductive properties and photosensitizing activity of PCSs. Alongside this, some PCSs reveal an effect of photoinitiated oxygen adsorption. However, it is to be emphasized that there is no direct correspondence between photosensitizing activity of PCSs and their ability for photoinitiated oxygen adsorption. On the contrary, as was demonstrated for the products of the thermal transformation of PAN, there is a reversed relationship between these characteristics, i. e., polymers revealing high photoactivated oxygen adsorption possess moderate photosensitizing activity, and vice versa. The highest photosensitizing activity is specific to polyquinolines which manifest no photoactivated oxygen adsorption at all. It can be suggested that the reactions of photosensitized oxidation of substrate and photoactivated oxygen adsorption by PCSs are accompanying the decomposition of the intermediate complex as competing processes. The rate of photosensitized oxidation of ascorbic acid by oxygen increases both with increased illumination intensity and increased energy of quantum of light.

On the other hand, some PCSs have demonstrated an effect which could be named "photoinhibition of the catalytic oxidation process". Thus, it can be seen from Fig. 22 that poly(propionitrile) catalyzing the ascorbic acid oxidation in darkness manifests suppressed catalytic activity on exposure to light.

---

[b] The intensity of the "white" light in the experiments was 3.7×10^{-3} and 2.6×10^{-2} J $cm^{-2}s^{-1}$ (3.7×10^4 and 2.6×10^5 erg $cm^{-2}s^{-1}$). The intensity of the "red" light was 2.5×10^{-3} J $cm^{-2}s^{-1}$ (2.5×10^4 erg $cm^{-2}s^{-1}$).

[c] The amount of oxygen adsorbed was referred to the unit surface of PCS samples.

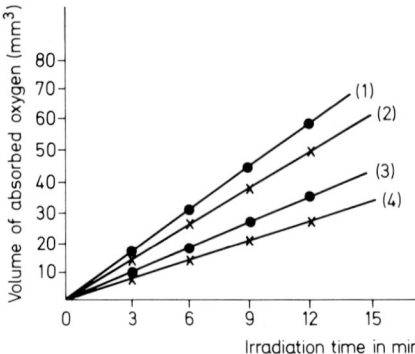

Fig. 22. Kinetic curves of oxygen absorption by ascorbic acid in the presence of poly(propionitrile). (1): in the dark; (2): in red light; (3), (4): in white light

A similar phenomenon has been observed for poly(acetonitrile) and poly(paracyanogen). This phenomenon is, apparently, due to a photodesorption[40] noted for these polymers and some inorganic semiconductor catalysts. According to the concepts on the origin of photodesorption developed by some researchers[269], it results from the lowering of electron concentration on the surface of the adsorbent, caused by irradiation or annihilation of the excitons at the adsorption centers accompanied by the change of the bond character between adsorbent and adsorbate.

5.2. Catalytic Activity

The catalytic activity of PCSs results from their semiconductor properties. The first studies in this field date from 1959–1961. Thus, we have demonstrated catalytic activity of products of the thermal transformation of PAN in the decomposition reactions of hydrogen peroxide, hydrazine hydrate, and formic acid[270, 271]. There is an indication of catalytic activity of poly(aminoquinone) in the reactions of the hydrogen peroxide decomposition[272].

PCSs obtained by dehydrochlorination of poly(2-chlorovinyl methyl ketones) catalyze the processes of oxidation and dehydrogenation of alcohols, and the toluene oxidation[207]. The products of the thermal transformation of PAN are also catalysts for the decomposition of nitrous oxide, for the dehydrogenation of alcohols and cyclohexene[274], and for the cis-trans isomerization of olefins[275]. Catalytic activity in the decomposition reactions of hydrazine, formic acid, and hydrogen peroxide is also manifested by the products of PVC dehydrochlorination[276, 277].

High catalytic activity is demonstrated by phthalocyanines in the reactions of the oxidation of benzaldehyde, oleic acid, and cyclohexene[278]. Copper, nickel, and cobalt phthalocyanines are active catalysts of cumene oxidation to hydroperoxide[279]. A detailed study of the catalytic activity of a number of polyconjugated chelate compounds in this reaction made it possible to elucidate the influence of the metal forming the chelate and of the structure of PCS[273, 280]. Reactions of deuterio exchange and hydrogen *ortho-para* conversion are also catalyzed by PCSs[281]. PCSs are active catalysts in the reactions of the dehydrogenation of higher alcohols[282, 283], of the oxidation of aryl aromatic hydrocarbons[284], and in some other reactions. At present the problem of the mechanism of the catalytic activity of PCSs cannot be considered as solved. It has been pointed out that some researchers approach this problem from the standpoint of semiconductor catalysis[285]. Others attach the decisive importance to the paramagnetism of PCSs[281, 286, 287]. Indeed, in a number of cases a distinct correspondence between the concentration of paramagnetic centers detected by ESR and the catalytic activity has been found. On the other hand, in certain cases the catalytic activity correlates to a greater degree with such characteristics of PCSs as electrical conductivity and activation energy of the conductivity[270, 271]. There are, evidently, reasons to suggest that structural correspondence of the catalytic system and the substrate is also of importance. Thus, it has been demonstrated that the introduction of a polymeric "seed" accelerates considerably the polymerization of pyridine and quinoline (with ring opening) only if this seed is itself a product of polymerization of the respective monomers. Otherwise, the catalytic activity does not manifest itself[288]. Apparently, autocatalysis in the process of the thermal transformation of PAN is of the same nature. In conclusion we shall point out that the concept of the decisive role of paramagnetic centers per se was taken as a basis for the hypothesis of the presence of the "local activation effect"[289]. The nature of this effect consists, in our opinion, in the presence of paramagnetic centers bringing about an increased reactivity and the change of chemical properties of diamagnetic components complexed with them.

The effect of local activation accounts for such features of PCSs as catalytic and stabilizing properties, autocatalysis, specific properties of thermal degradation, structural modification, and a number of other phenomena typical of polymers with a system of conjugated bonds.

6. References

[1] Volkenstein, M. V.: Configurational statistics of polymer chains. New York: Interscience 1963
[2] Mylnikov, V. S.: Usp. Khim. **43**, 1821 (1974); C.A. **82**, 58128 (1975)
[3] Vymazal, Z., Stepek, J.: Chem. Listy **59**, 800 (1965)
[4] Kargin, V. A., Topchiev, A. V., Krentsel, B. A., Polak, L. S., Davydov, B. E.: Zh. Vses. Khim. Ova. **5**, 507 (1960); C.A. **55**, 5109 (1961)
[5] Dolinski, R. J., Dean, W. R.: Chem. Technol. **1971**, 304; Chim. Ind. (Milan) **53**, 952 (1971)
[6] Mark, H.: Israel J. Chem. **10**, 407 (1972)
[7] May, J. F.: Chim. Ind., Genie Chim. **104**, 1243 (1971)
[8] Frumkin, A. N., Davydov, B. E.: Vestn. Akad. Nauk SSR **10**, 33 (1971)
[9] Paushkin, Ya. M., Vishnyakova, T. P., Lunin, A. F., Nizova, S. A.: Organicheskie Polimernye Poluprovodniki Moscow: Khimiya, 1971; C.A. **77**, 6663 (1972)
[10] Kargin, V. A. (ed.): Organicheskiya poluprovodniki (organic semiconductors) Moscow: Nauka 1968
[11] Brophy, J. J., Butterey, N. (eds.): Organic semiconductors. New York: Collitor 1962; Boguslavskii, L. I., Vannikov, A. V.: Organicheskiya poluprovodniki i biopolimery (organic semiconductors and biopolymers). Moscow: Nauka, 1963; Kryszewski, M.: Polprzewodniki wielkoczasteczcowe. Warsaw: Panstwowe Wydawnictwo Naukowe, 1968; Jnokuchi, H., Akamatu, R.: Electrical conductivity of organic Semiconductors. New York – London: Academic, 1961
[12] Natta, G., Pino, P., Mazzanti, G.: Ital. Patent 530753 (1955); C.A. **52**, 15128 (1958)
[13] Noguchi, H., Kambara, S.: J. Polym. Sci., Part B, **1**, 553 (1963)
[14] Watson, W. H., McMordie, W. C., Lands, L. G.: J. Polym. Sci. **55**, 137 (1961)
[15] Luttinger, L. B.: Chem. Ind. (London) **1960**, 1135
[16] Luttinger, L. B.: J. Org. Chem. **27**, 1591 (1962)
[17] Tabata, Y., Saito, B., Shibano, H., Sobue, H., Oshima, K.: Makromol. Chem. **76**, 89 (1964)
[18] Kambara, S., Hatano, M., Sakaguchi, K.: J. Polym. Sci. **51**, S7 (1961)
[19] Bantsyrev, G. J., Cherkashin, M. J., Kalikhman, J. D.: XVII Conference of High Molecular Compounds, USSR, Collection of Abstracts, Moscow 1969, p. 122
[20] Shantarovich, P. S., Shlyapnikova, J. A.: Vysokomol. Soedin. **3**, 363 (1961); C.A. **55**, 26517 (1961)
[21] Byrd, N. R., Kleist, F. D., Rembaum, A.: J. Macromol. Sci., Chem. **A-1**, 627 (1967)
[22] Kambara, S., Noguchi, H.: Makromol. Chem. **73**, 244, (1964)
[23] Barcalov, J. M., Berlin, A. A., Gol'danskii, V. J., Go-Min-gao: Vysokomol. Soedin. **5**, 368 (1963); C.A. **59**, 763 (1963)
[24] Berlin, A. A., Cherkashin, M. J., Kisilitsa, P. P., Pirogov, O. N.: Vysokomol. Soedin, Ser. A, **9**, 1835 (1967); C.A. **67**, 109027 (1967)
[25] Natta, G., Mazzanti, G., Pregaglia, G., Peraldo, M.: Gazz. Chim. Ital. **89**, 465 (1959)
[26] Trepka, W. J., Sonnenfeld, R. J.: J. Polym. Sci., Part A-1, **8**, 2721 (1970)
[27] Barkalov, J. M., Gol'danskii, V. J., Dzantiev, B. G., Kuzmina, S. S.: Proceedings of the II USSR conference of radiation chemistry, 455 (1962); C.A. **58**, 4652 (1963)
[28] Berlin, A. A., Cherkashin, M. J., Chernysheva, J. P., Aseev, Yu. G., Barkan, E. J., Kisilitsa, P. P.: Vysokomol. Soedin., Ser. **A-9**, 1840 (1967); C.A. **67**, 109090 (1967)
[29] Kriz, J., Benes, M. J., Peska, J.: Collect. Czech. Chem. Commun. **32**, 4043 (1967)
[30] Sobue, H., Tabata, Y.: Proceedings of the Symposium on Polymers, Tokyo 1961
[31] Peska, J., Benes, M. J.: Collect. Czech. Chem. Commun. **38**, 2595 (1973)
[32] Benes, M. J., Peska, J., Wichezel, O.: Materialy Conferenzii of High Molecular Compounds, Paris 1963
[33] Simionescu, Cr., Dumitrescu, Sv., Lixandru, T., Vasiliu, S., Simionescu, B.: Plaste Kautsch. **19**, 101 (1972)
[34] Janousova, A., Benes, M. J., Janic, M., Peska, J.: Collect. Czech. Chem. Commun. **39**, 1858 (1974); Collect. Czech. Chem. Commun. **38**, 3762 (1973).

35) Simionescu, Cr., Dumitrescu, Sv., Perces, V.: Plaste Kautsch. **20**, 913 (1973)
36) Nasirov, F. M., Krentsel, B. A., Davydov, B. E.: Izv. Akad. Nauk SSSR, Ser. Khim. **1965** 1009; C.A. **63**, 10071 (1965)
37) Davydov, B. E., Demidova, G. N., Pirtskhalawa, R. N., Rosenshtein, Z. D.: Elektrokhimiya **1**, 876 (1965); C. A. **63**, 14988 (1965)
38) Nasirov, F. M., Karpacheva, G. P., Davydov, B. E., Krentsel, B. A. Izv. Akad. Nauk SSSR., Ser. Khim. **9**, 1697 (1964); C.A. **62**, 217 (1965)
39) Nasirov, F. M.: Dissertation for the Degree of candidate of Sciences, Institute of Petrochemical Synthesis, USSR, Academy of Sciences, Moscow 1965
40) Davydov, B. E.: Dissertation for the Degree of Doctor of Sciences, Institute of Petrochemical Synthesis, USSR Academy of Sciences, Moscow 1966
41) Berlin, A. A., Geiderikh, M. A., Davydov, B. E., Kargin, V. A., Karpacheva, G. P., Krentsel, B. A., Khutareva, G. V.: Khimiya polysopryazhennykh sistem (chemistry of polyconjugated systems). Moscow: Khimiya, 1972
42) Davydov, B. E., Krentsel, B. A., Khutareva, G. V.: J. Polym. Sci., Part C, **16**, 1365 (1967)
43) Khutareva, G. V., Shishkina, M. V., Davydov, B. E.: Izv. Akad. Nauk SSSR, Ser. Khim., **3**, 520 (1965); C.A. **63**, 1877 (1965)
44) Khutareva, G. V., Krentsel, B. A., Shishkina, M. V., Davydov, B. E.: Neftekhimiya **5**, 90 (1965); C.A. **62**, 14835 (1965)
45) Khutareva, G. V.: Dissertation for the Degree of Candidate of Sciences, Institute of Petrochemical Synthesis, USSR Academy of Sciences, Moscow 1965
46) Geiderikh, M. A., Davydov, B. E., Zaliznaya, N. F., Oreshkina, G. A.: Vysokomol. Soedin., Ser. B, **11**, 870 (1969); C.A. **72**, 90904 (1970)
47) Zaliznaya, N. F., Collection of proceedings of the conference of young researchers, Institute of Petrochemical Synthesis, USSR Academy of Sciences, Moscow 1973
48) Zaliznaya, N. F., Geiderikh, M. A., Davydov, B. E., Zemtsov, L. M.: Abstracts of the XVIII conference on high molecular compounds, Kazan, 1973; Moscow: Nauka, 1973, p. 38
49) Serebrynicov, V. S., Davydov, B. E., Stotskaya, L. L., Krentsel, B. A.: USSR Inventor's certificate, 368801, 30.06.70., Bulletin No. 10, 1973
50) Serebryanicov, V. S., Davydov, B. E., Karpacheva, G. P., Stotskaya, L. L.: Preprints of the XX International Symposium on Macromolecular Chemistry, Madrid 1974, Vol. 1, p. 396
51) Ellinger, L. P.: Polymer **6**, 549 (1965)
52) Sitnikova, T. A., Stotskaya, L. L., Shimanko, N. A., Krentsel, B. A.: Dokl. Akad. Nauk SSSR **181**, 401 (1968), C.A. **69**, 87537 (1968); ibid.: **207**, 878 (1972)
53) Stotskaya, L. L., Sitnikova, T. A., Krentsel, B. A.: Europ. Polym. J. **7**, 1661 (1971)
54) Polyakova, A. M., Korshak, V. V., Suchkova, M. D.: Vysokomol. Soedin. **4**, 486 (1962); C.A. **58**, 12682 (1963)
55) Kambara, S.: Proceedings of the Symposium on Polymers, Tokyo 1961
56) Kargin, V. A., Kabanov, V. A., Zubov, V. P., Zezin, A. B.: Dokl. Akad. Nauk SSSR **139**, 605 (1961); C.A. **56**, 2569 (1962)
57) Kabanov, V. A., Zubov, V. P.: Zh. Vses. Khim. Ova. **7**, 131 (1962); C.A. **57**, 4813 (1962)
58) Kabanov, V. A., Kargin, V. A., Kovaleva, V. A., Topchiev, D. A.: Vysokomol. Soedin. **6**, 1852 (1964); C.A. **62**, 5342 (1965); Kabanov, V. A., Topchiev, D. A., Lebedeva, T. S., Popov, V. G., Kargin, V. A.: Vysokomol. Soedin., Ser. A, **10**, 1762 (1968); C.A. **69**, 77792 (1968)
59) Kabanov, V. A., Zubov, V. P., Kovaleva, V. P., Kargin, V. A.: J. Polym. Sci., Part C, **4**, 1009 (1964)
60) Topchiev, D. A., Kabanov, V. A., Kargin, V. A.: Vysokomol. Soedin. **6**, 1814 (1964); C.A. **62**, 6571 (1965)
61) Topchiev, D. A., Popov, V. G., Kabanov, V. A., Kargin, V. A.: Izv. Akad. Nauk SSSR., Ser. Khim., **1964**, 391; C.A. **60**, 13327 (1964)
62) Topchiev, D. A., Popov, V. G., Shishkina, M. V., Kabanov, V. A., Kargin, V. A.: Vysokomol. Soedin. **8**, 1767 (1966); C.A. **66**, 11196 (1967)
63) Topchiev, D. A.: Dissertation for the Degree of Candidate of Sciences, Institute of Petrochemical Synthesis, USSR Academy of Sciences, Moscow 1965

64) Hay, A. S.: J. Org. Chem. 25, 1275 (1960)
65) Korshak, V. V., Sladkov, A. M., Kudryavtsev, Yu. P.: Vysokomol. Soedin. 2, 1824 (1960); C.A. 55, 26522 (1961)
66) Kotlyarevsky, J. L., Fisher, L. B., Dulov, A. A., Slinkin, A. A.: Izv. Akad. Nauk SSSR, Otd. Khim. Nauk 1960, 950; C.A. 54, 24466 (1960)
67) Berlin, A. A.: Proceedings of the International Symposium on Macromolecular Chemistry, Moscow 1960
68) Parini, V. P., Kazakova, Z. S., Okorokova, M. N., Berlin, A. A.: Vysokomol. Soedin. 3, 402 (1961); C.A. 55, 27956 (1961)
69) Kavacic, P., Kyriakis, A.: Tetrahedron Lett. 1962, 467
70) Kovacic, P., Koch, F. W., Stephan, G. E.: J. Polym. Sci., Part. A, 2, 1193 (1964)
71) Bilow, N., Miller, L. J.: J. Macromol. Sci., Chem. 1, 183 (1967)
72) Parini, V. P., Kazakova, Z. S., Berlin, A. A.: Vysokomol. Soedin. 3, 1870 (1961); C.A. 56, 14460 (1962)
73) Constantini, P., Belorgey, G., Jozefowicz, M., Buvet, R.: C.R. Hebd. Seances Acad. Sci. 258, 6421 (1964)
74) Jozefowicz, M., Belorgey, G., Yu, L. T., Buvet, R.: C.R. Hebd. Seances Acad. Sci. 260, 6367 (1965)
75) Jozefowicz, M., Yu, L. T.: Rev. Gen. Electr. 75, 1008 (1966)
76) Kotlyarevskii, J. L., Terpugova, M. P., Andrievskaya, E. K.: Izv. Akad. Nauk SSSR, Ser. Khim., 1964, 1854
77) Bach, H. C., Biack, W. B.: Internat. Symposium on Macromolecular Chemistry in Brussels, Prepr., 4/48 (1967); Polym. Prepr. Am. Chem. Soc., Div. Polym. Chem. 7, 576 (1966)
78) Hay, A. S., Blanchard, N. S., Endres, G. F., Eustance, J. W.: J. Am. Chem. Soc. 81, 6335 (1959)
79) Staffin, G. D., Price, C. C.: J. Am. Chem. Soc. 82, 3632 (1960)
80) Berlin, A. A., Liogon'kii, B. J., Zelenetskii, A. N.: Izv. Akad. Nauk SSSR, Ser. Khim., 1967, 225; The Internat. Symposium on Macromol. Chem., Brussels, Preprint (1967)
81) Berlin, A. A., Liogon'kii, B. J., Zelenetskii, A. N.: Vysokomol. Soedin. Ser. A, 10, 2089 (1968); C.A. 70, 12048 (1969); Dokl. Akad. Nauk SSSR 178, 1320 (1968); C.A. 68, 115039 (1968); J. Polym. Sci. Part C, 22, 443 (1968)
82) Schopov, J.: J. Polym. Sci., Polym. Lett. Ed. 4, 1023 (1966)
83) Goldfinger, G.: J. Polym. Sci. 4, 93 (1949); Edwards, G. A., Goldfinger, G.: J. Polym. Sci. 16, 589 (1955)
84) Berlin, A. A., Vonsiatskii, V. A., Liogon'kii, B. J.: Dokl. Akad. Nauk SSSR 144, 1316 (1962); C.A. 57, 16862 (1962)
85) Slonim, J. Ya., Urman, Ya. G., Vonsiatskii, V. A., Liogon'kii, B. J., Berlin, A. A.: Dokl. Akad. Nauk SSSR 154, 914 (1964); C.A. 60, 14627 (1964)
86) Bloomfild, P. R., Parvin, K.: Brit. Pat. 928576 (1963); C.A. 59, 10303 (1963)
87) Barcel, J., Rene, B.: C. R. Hebd. Seances Acad. Sci. 252, 1801 (1961)
88) Jozefowicz, M.: Bull. Soc. Chim. Fr. 1963, 2036
89) Heitz, W., Ullrich, R.: Makromol.Chem. 98, 29 (1966)
90) Kern, E., Gehm, R.: Angew. Chem. 62, 337 (1950); Acta Chem. Scand. 5, 270 (1951)
91) Kern, W., Wirth, O. H.: Kunststoffe-Plastics 6, 12 (1958); Makromol. Chem. 29, 164 (1959)
92) Wirth, H. O., Müller, R., Kern, W.: Makromol. Chem. 77, 90 (1964)
93) Berlin, A. A., Matveeva, N. G.: Izv. Akad. Nauk SSSR, Otd. Khim. Nauk. 1959, 2260
94) Macallum, A. D.: J. Org. Chem. 13, 154 (1948), USA Pat. 2513188 (1950); USA Pat. 2538941 (1951)
95) Lenz, R. W., Carrington, W. K.: J. Polym. Sci. 41, 333 (1959)
96) Lenz, R. W., Handlovits, C. E.: J. Polym. Sci. 43, 167 (1962)
97) Lenz, R. W., Handlovits, C. E., Smith, H. A.: J. Polym. Sci. 58, 351 (1962)
98) Berlin, A. A., Matveeva, N. G.: Vysokomol. Soedin. 1, 1643 (1959); C.A. 54, 16899 (1960)
99) Berlin, A. A., Liogon'kii, B. L., Parini, V. P.: Vysokomol. Soedin. 2, 689 (1960); C.A. 55, 6907 (1961); Vysokom. Soedin. 3, 1491 (1961); C.A. 56, 10387 (1962); Izv. Akad. Nauk SSSR, 1964, 705; C.A. 61, 5797 (1964); Proceedings of the International Symposium on Macromolecular Chemistry, Doklady, Sektsiya 3, p. 115; C.A. 56, 395 (1962)

100) Berlin, A. A., Liogon'kii, B. J., Parini, V. P., Leikina, M. S.: Vysokomol. Soedin. **4**, 662 (1962); C.A. **58**, 11475 (1963)
101) Carlton, D. M., McCarthy, D. K., Genz, R. H.: J. Phys. Chem. **68**, 2661 (1964)
102) Poninski, M., Kryszewski, M.: Roczniki Chem. **39**, 505 (1965); C.A. **63**, 10083 (1965)
103) Berlin, A. A., Gudvilovich, T. V., Parini, V. P., Sorokin, A. V.: Izv. Akad. Nauk SSSR, Ser. Khim. **1966** 2038; C.A. **66**, 80603 (1967); Parini, V. P., Gudvilovich. J. V.: Izv. Akad. Nauk SSSR, Ser. Khim. **1965**, 370; C.A. **62**, 14550(1965)
104) Ravve, A., Fitko, C.: J. Polym. Sci., Part A, **2**, 1925 (1964)
105) Berlin, A. A., Belova, G. V., Gudvilovich. J. V.: Vysokomol. Soedin. Ser. A **9**, 2214 (1967); C.A. **68**, 3232 (1968)
106) Berlin, A. A., Razvodovskii, E. F.: Dokl. Akad. Nauk SSSR **140**, 598 (1961); C.A. **56**, 2562 (1962); J. Polym. Sci., Part. C, **16**, 369 (1967)
107) Berlin, A. A., Zherebtsova, L. V., Razvodovskii, E. F.: Vysokomol. Soedin. **6**, 58 (1964); C.A. **60**, 10800 (1964)
108) Berlin, A. A., Matveeva, N. G.: Usp. Khim. **29**, 277 (1970); C.A. **54**, 16370 (1960)
109) Korshak, V. V.: Usp. Khim. **35**, 1030 (1966), C.A. **65**, 10670 (1966)
110) Terent'ev, A. P., Rode, V. V., Rukhadze, E. G., Vozhennikov, V. M., Zvonkova, Z. V., Badzhadze, L. M.: Dokl. Akad. Nauk SSSR **140**, 1093 (1961)
111) Rode, V. V., Rukhadze, E. G., Terent'ev, A. P.: Usp. Khim. **32**, 1488 (1963); C.A. **60**, 9357 (1964)
112) Joiner, R. D., Kenney, M. E.: J. Am. Chem. Soc. **82**, 5791 (1960)
113) Korshak, V. V., Vinogradova, S. V., Morozova, D. T.: Vysokomol. Soedin. **3**, 1500 (1961); C.A. **56**, 13080 (1962)
114) Kanda, S., Saito, Y.: Bull. Chem. Soc. Jpn. **30**, 192 (1957)
115) Zimmerman, B., Lochte, H.: J. Am. Chem. Soc. **60**, 2456 (1938)
116) Marvel, C. S., Hill, H. W.: J. Am. Chem. Soc. **72**, 4819 (1950). Marvel, C. S., Tarkoy, N.: J. Am. Chem. Soc. **79**, 6000 (1957): ibid., **80**, 832 (1958); Marvel, C. S., Bonsignore, P. V.: J. Am. Chem. Soc. **81**, 2668 (1959)
117) Korshak, Yu. V.: Dissertation for the Degree of Candidate of Sciences, Institute of Petrochemical Synthesis, USSR Academy of Sciences, Moscow 1964
118) Topchiev, A. V., Korshak, Yu. V., Popov, Yu., A., Rosenstein, L. D.: J. Polym. Sci., Part C, **4**, 1305 (1963)
119) Topchiev, A. V., Korshak, Yu. V., Davydov, B. E., Krentsel, B. A.: Dokl. Akad. Nauk, SSSR **147**, 645 (1962); C.A. **58**, 8047 (1963)
120) Korshak, Yu. V., Pronyik, T. A., Davydov, B. E.: Neftekhimiya **3**, 677 (1963); C.A. **60**, 3108 (1964)
121) Lapitskii, G. A., Makin, S. M.: Zh. Vses. Khim. Ova. **9**, 116 (1964); C.A. **60**, 14418 (1964)
122) Lapitskii, G. A.: Dissertation for the Degree of Candidate of Scieces, Moscow Lomonosov Institute of Fine Chemical Technology, 1965
123) Cordes, E. H., Jencks, W. P.: J. Am. Chem. Soc. **84**, 832 (1962)
124) Krässig, H., Greber, G.: Makromol. Chem. **17**, 131 (1956); Makromol. Chem. **17**, 154 (1956)
125) Bayer, E., Chem. Ber. **90**, 2785 (1957)
126) Popov, Yu. A.: Dissertation for the Degree of Candidate of Sciences, Institute of Petrochemical Synthesis, USSR Academy of Sciencas, Moscow 1963
127) Popov, Yu. A., Davidov, B. E., Shishkina, M. V., Krentsel, B. A.: Izv. Akad. Nauk SSSR, Ser. Khim. **1963**, 2014; C.A. **60**, 9368 (1964)
128) Davydov, B. E., Popov, Yu. A., Prokof'eva, L. V., Rostenshtein, L. D.: Izv. Akad. Nauk SSSR. Ser. Khim. **1963**, 759; C.A. **59**, 10843 (1963)
129) Raskina, E. M., Perekal'skaya, L. M., Davydov; B. E., Shishkina, M. V.: Elektrokhimiya **2**, 1332 (1966)
130) D'Alelio, G. F., Hornback, J. M., Strazik, W. F., Schoenig, R. K., Kurosaki, T.: J. Makromol. Sci., Chem. **2**, 237 285, 335, 474 (1968); J. Macromol. Sci., Chem. **1**, 1251, 1261, 1299, 1321, 1331 (1967)
131) Berlin, A. A., Parini, V. P.: Izv. Akad. Nauk SSSR, Ser. Khim. **1965**, 204; C.A. **62**, 13101 (1965)

132) Kuhn, H.: Angew. Chem. 71, 93 (1959)
133) Kargin, V. A., Kabanov, V. A., Zubov, V. P., Papisov, J. M., Kurochkina, G. J.: Dokl. Akad. Nauk SSSR 140, 122 (1961); C.A. 56, 2566 (1962)
134) Kudryavtsev, G. J., Vasil'eva-Sokolova, E. A., Mazel, J. S.: Vysokomol. Soedin. 5, 151 (1963); C.A. 59, 770 (1963)
135) Pen'kovskii, V. V.: Vysokomol. Soedin. 6, 1755 (1964); C.A. 62, 2833 (1965)
136) McDonald, R. N., Campbell, T. W.: J. Am. Chem. Soc. 82, 4669 (1960)
137) Yanovskaya, L. A.: Usp. Khim. 30, 813 (1961); C.A. 56, 1323 (1962)
138) Bircumshaw, L. L., Tayler, F. M., Whiffen, D. H.: J. Chem. Soc. 1954, 931
139) Davydov, B. E., Raskina, E. M., Krentsel, B. A.: Vysokomol. Soedin. 4, 1604 (1962); C.A. 58, 14125 (1963)
140) Wrazidlo, W., Levine, H. H.: J. Polym. Sci., Part A, 2, 4795 (1964)
141) Marvel, C. S.: SPE. J. 20, 220 (1964); C.A. 60, 15985 (1964)
142) Korshak, V. V., Krongauz, E. S.: Usp. Khim. 33, 1409 (1964); C.A. 62, 9237 (1965)
143) Doroshenko, Yu. E.: Usp. Khim. 36, 1346 (1967); C.A. 67, 117288 (1967)
144) Hergenrother, P. M., Wrazidlo, W., Levine, H. H.: J. Polym. Sci., Part A, 3, 1665 (1965)
145) Bower, G. M., Frost, L. M.: J. Polym. Sci. Part A, 1, 3135 (1963)
146) Van Deusen, R. L.: J. Polym. Sci., Part B, 4, 211 (1966)
147) Dawans, F., Marvel, C. S.: J. Polym. Sci., Part A, 3, 3549 (1965)
148) Bell, V. L., Perdirtz, G. F.: J. Polym. Sci., Part B, 3, 977 (1965)
149) Colson, J. G., Michel, R. H., Paufler, R. M.: J. Polym. Sci., Part A, 4, 59 (1966)
150) Berlin, A. A., Liogon'kii, B. J. Shamraev, G. M., Belova, G. V.: Izv. Akad. Nauk SSSR., Ser. Khim. 1966, 945; C.A. 65, 13874 (1966); Vysokomol. Soedin., Ser. A, 9, 1936 (1967); C.A. 67, 109042 (1967)
151) Huisgen, R.: Angew. Chem. 75, 604, 742 (1963)
152) Korshak, V. V., Krongauz, E. S., Berlin, A. M.: Dokl. Akad. Nauk SSSR 152, 1108 (1963); C.A. 60, 2922 (1964)
153) Berlin, A. M.: Dissertation for the Degree of Candidate of Sciences, Institute of Element – Organic Chemistry, USSR Academy of Sciences 1966
154) Overberger, C. G., Fujimoto, S.: J. Polym. Sci., Part B, 3, 735 (1965)
155) Jwakura, J., Akiyama, M., Shiraishi, S.: Bull. Chem. Soc. Jpn. 38, 335 (1965)
156) Akiyama, M., Jwakura, Y., Shiraishi, S., Jmai, Y.: J. Polym. Sci., Part B, 4, 305 (1966)
157) Berlin, A. A., Grigorovskaya, V. A., Skurat, V. E.: Vysokomol. Soedin. 8, 1976 (1966)
158) Sukhorukov, B. J., Kuz'menko, V. A., Blumenfeld, L. A. in collected articles: Geterotsepnyl vysokomolekularniye soedineniya (heterochain high-molecular compounds), 1964
159) Kryazhev, Yu. G., Baiborodina, E. N., Ermakova, T. G., Tatarova, Z. A., Brodskaya, E. J., Kalikhman, J. D., Salaurov, V. N., Myachin, Yu. A.: Vysokomol. Soedin., Part A, 16, 2272 (1974)
160) Paushkin, Ya. M., Omarov, O. Yu.: Vysokomol. Soedin. 7, 710 (1965); C.A. 63, 3056 (1965)
161) Kryazhev, Yu. G., Ermakova, T. G.: Vysokomol. Soedin., Ser. A, 15, 478(1973); C.A. 79, 42880 (1973)
162) Kryazhev, Yu. G., Brodskaya, E. J., Salaurov, V. N., Shibanova, E. F.: Vysokomol. Soedin., Ser. A, 14, 2360 (1972); C.A. 78, 98082 (1973)
163) Rath, H., Heise, L.: Kunststoffe 44, 8, 341 (1954)
164) Korshak, V. V., Samplavskaya, K. K., Dovol'skaya, J. M., Zh. Obsch. Khim. 20, 2080 (1950)
165) Braun, D., Thallmaier, M.: Makromol. Chem. 99, 59 (1966)
166) Marvel, C. S., Sample, J. H., Roy, M. E.: J. Am. Chem. Soc. 61, 3241 (1939)
167) Marvel, C. S., Sample, J. H., Roy, M. E.: J. Am. Chem. Soc. 64, 2356 (1942); J. Am. Chem. Soc. 65, 1716 (1943)
168) Flory, P. J.: J. Am. Chem. Soc. 61, 1581 (1939)
169) Bohrer, J.: Trans. N. J. Acad. Sci. 2, 20 (1955); Trans. N. J. Acad. Sci. 5, 367 (1958)
170) Kudryavtsev, Yu. P., Sladkov, A. M., Aseev, Yu. G., Nedoshivin, Yu. N., Kasatochkin, V. J., Korshak, V. V.: Dokl. Akad. Nauk SSSR 158, 389 (1964); C.A. 62, 12608 (1965)

171) Tokarzewski, L.: Roczniki, Chem. **33**, 619 (1959)
172) Tokarzewski, L.: Roczniki Chem. **33**, 849 (1959)
173) Astaf'ev, J. V., Piskunov, A. K.: Vysokomol. Soedin. **2**, 1745 (1960)
174) Adylov, S. A., Il'ina, D. E., Krentsel, B. A., Shishkina, M. V.: Vysokomol. Soedin. **5**, 316 (1963)
175) Matheson, L. A., Boyer, R. F., Ind. Energ. Chem. **44**, 867 (1952)
176) Fuchs, N.: Makromol. Chem. **22**, 1 (1957)
177) Korshak, V. V., Zamyatina, V. A.: Zh. Prikl. Khim. (Leningrad) **14**, 809 (1941)
178) Volkober, Z.: Proceedlings of the Symposium on Heat-Resistant Polymer, London, 1960
179) Bevington, J.: J. Chem. Soc. **1940**, 771
180) Berlin, A. A., Aseeva, R. M., Kalyaev, G. I., Frankevich, E. L.: Dokl. Akad. Nauk SSSR **144**, 1042 (1962); C.A. **57**, 15331 (1962)
181) Berlin, A. A., Kasatochkin, V. J., Aseeva, R. M., Finkel'shtein, G. B.: Vysokomol. Soedin. **5**, 1303 (1963); C.A. **60**, 667 (1964)
182) Berlin, A. A., Aseeva, R. M., Smutkina, L. S., Kasatochkin, V. J.: Izv. Akad. Nauk SSSR, Ser. Khim. **1965**, 1974; C.A. **62**, 9254 (1965)
183) Ladd, E. C., Shinkle, S. D.: USA Pat. 2561516; C.A. **46**, 2561 (1952)
184) Krentsel, B. A., Semenido, G. E., Il'ina, D. E.: Vysokomol. Soedin. **5**, 558 (1963); C.A. **59**, 1773 (1963)
185) Land, E. H.: USA Pat. 2305108 (1942)
186) Mainthia, S. B., Kronick, P. L., Labes, M. M.: J. Chem. Phys. **37**, 2509 (1962)
187) Burnett, G. M., Haldon, R. A., Hay, J. N.: Eur. Polym. J. **3**, 449 (1967)
188) Grassie, N.: Chemistry of high polymer degradation processes. London: Butterworths 1956
189) Guyot, A., Benevise, J., Trambouse, Y.: J. Appl. Polym. Sci. **6**, (19) 103 (1962)
190) Rieche, A., Grimm, A., Mucke, H.: Kunststoffe **52**, 265 (1962)
191) Talamini, G., Cinque, G., Palma, G.: Mater. Plast. Elastomeri **30**, 317 (1964)
192) Troitskaya, L. S., Myakov, V. N., Troitskii, B. B., Rasuvaev, G. A.: Vysokomol. Soedin., Ser., A, **9**, 2119 (1967); C.A. **68**, 3221 (1968)
193) Mikhailov, N. V., Tokareva, L. G., Klemenkov, V. S.: Kolloidn. Zh. **18**, 578, 597 (1956)
194) Winkler, D. E.: J. Polym. Sci., **35**, 3 (1959)
195) Aseeva, R. M., Aseev, Yu. G., Berlin, A. A., Kasatochkin, V. J.: Zh. Strukt. Khim. **6**, 47 (1965); C.A. **63**, 1886 (1965)
196) Campbell, J. E., Rausher, W. H.: J. Polym. Sci. **18**, 461 (1955)
197) Marvel, C. S., Hartzell, G. E.: J. Am. Chem. Soc. **81**, 448 (1959)
198) Marvel, C. S., Vest, R. D.: J. Am. Chem. Soc. **79**, 5771 (1957)
199) Stefanovskaya, N. N., Gavrilenko, J. F., Markevich, J. N., Shmonina, V. L., Tinyakova, E. J., Dolgoplosk, B. A.: Izv. Akad. Nauk SSSR, Ser. Khim. **1967**, 2355; C.A. **68**, 13637 (1968)
200) Stefanovskaya, N. N., Shmonina, V. L., Gavrilenko, J. F., Tinyakova, E. I., Dolgoplosk, B. A.: Dokl. Akad. Nauk SSSR **174**, 1356 (1967); C.A. **68**, 13836 (1968)
201) Stefanovskaya, N. N., Gavrilenko, J. F., Tinyakova, E. I., Dolgoplosk, B. A.: Dokl. Akad. Nauk SSSR **175**, 95 (1967)
202) Markevich, J. N., Beilin, S. J., Teterina, M. P., Karpacheva, G. P., Dolgoplosk, B. A.: Dokl. Akad. Nauk SSSR **191**, 362 (1970); C.A. **73**, 15465 (1970)
203) Samedova, T. G., Markevich, J. N., Beilin, S. J., Davydov, B. E.; Vysokomol. Soedin., Ser. A. **15**, 2204 (1973)
204) Marvel, C. S.: J. Am. Chem. Soc., **60**, 280 (1938), J. Am. Chem. Soc., **61**, 3224 (1939)
205) Grassie, N., Hay, T. N.: Makromol. Chem. **64**, 82 (1963)
206) Nesmeyanov, A. N., Rubinshtein, A. M., Dulov, A. A., Slinkin, A. A., Rybinskaya, M. J., Slonimskii, G. L.: Dokl. Akad. Nauk SSSR. **135**, 609 (1960); C.A. **55**, 15338 (1961)
207) Nesmeyanov, A. N., Rubinskaya, M. J., Slonimskii, G. L.: Vysokomol. Soedin. **2**, 526 (1960); C.A. **55**, 4408 (1961)
208) Distler, G. J., Sotnikov, P. S., Kortukova, E. J.: Dokl. Akad. Nauk SSSR. **156**, 652 (1964); C.A. **61**, 5848 (1964)
209) Shindo, A., Soma, J.: Symposium on Carbon, Tokyo, p. 20 (1964)

210) Grassie, N.: Trans. Faraday Soc. **48**, 379 (1952)
211) Dole, M., Milner, D. C., Williams, T. F.: J. Am. Chem. Soc. **79**, 4809 (1957)
212) Bakh, N. A., Vannikov, A. V., Grishina, A. D., Markova, Z. A., Nizhnii, S. V. in: Radiatsionnaya khimiya polimerov (radiation chemistry of polymers) Moscow: Nauka, 1966, p. 263
213) Markova, Z. A., Ershov, B. G., Bakh, N. A.: Vysokomol. Soedin. **6**, 131 (1964); C.A. **60**, 10818 (1964)
214) Andrianov, K. A., Asnovich, E. Z., Petrashko, A. J.: Khimiya bol'shikh molekul (Chemistry of big molecules). Moscow: Znanie 1961
215) Bakh, N. A., Bityukov, V. D., Vannikov, A. V., Grishina, A. D.: Dokl. Akad. Nauk SSSR **144**, 135 (1962); C.A. **57**, 10022 (1962)
216) Bakh, N. A., Vannikov, A. V., Grishina, A. D., Markova, Z. A., Nizhnii, S. V. in: Radiationnaya khimiya polimerov (radiation chemistry of polymers). Moscow: Nauka 1966, p. 249
217) Baum, B., Wastman, J. H.: J. Polym. Sci. **28**, 527 (1958)
218) Houtz, R. C.: Text. Res. J. **20**, 786 (1950)
219) Bircumshaw, L. L., Taylor, F. M.: J. Chem. Soc. **1954**, 931
220) Grassie, N., McNeill, J. K.: J. Chem. Soc. **1956**, 3929
221) Grassie, N., McNeill, J. C.: J. Polym. Sci. **27**, 207 (1958)
222) Grassie, N.: Materials of International Symposium on Macromolecular Chemistry, Moscow 1960
223) Tiemann, F.: Ber. Dtsch. Chem. Ges. **19**, 1475 (1886)
224) Schulz, J.: J. Polym. Sci. **28**, 438 (1958)
225) Grassie, N., McNeill, J. C.: J. Polym. Sci. **30**, 37 (1958); J. Polym. Sci. **39**, 211 (1959)
226) McCartney, Z. R.: Nat. Bur. Stand. (US), Circ. **525**, 123 (1953)
227) Beaman, R. G.: J. Am. Chem. Soc. **70**, 3115 (1948)
228) Oversberger, C. G., Pearce, E. M., Mayers, N.: J. Polym. Sci. **34**, 109 (1959)
229) Takata, T.: Bull. Chem. Soc. Jpn. **35**, 1438 (1962)
230) Takata, T.: Bull. Chem. Soc. Jpn. **37**, 1567 (1964)
231) Pimenov, G. G., Maklakov, A. J., Geiderikh, M. A., Davydov, B. E.: Vysokomol. Soedin., Ser. B, **9**, 535 (1967)
232) Geiderikh, M. A.: Dissertation for the Degree of Candidate of Sciences, Institute of Petrochemical Synthesis, USSR Academy of sciences, Moscow 1965
233) Kubasova, N. A., Shiskina, M. V., Zaliznaya, N. F., Geiderikh, M. A.: Vysokomol. Soedin., Ser. A, **10**, 1324 (1968); C.A. **69**, 36523 (1968)
234) Drabkin, J. A., Rosenshtein, L. D., Geiderikh, M. A., Davydov, B. E.: Dokl. Akad. Nauk SSSR. **154**, 197 (1964); C.A. **60**, 13389 (1964)
235) Geiderikh, M. A., Davydov, B. E., Krentsel, B. A., Kustanovich, J. M., Polak, L. C., Topchiev, A. V., Voitenko, P. M.: J. Polym. Sci. **54**, 621 (1961)
236) Oreshkina, G. A., Karpacheva, G. P., Geiderikh, M. A., Davydov, B. E.: Electrokhimiya **4**, 217 (1968); C.A. **68**, 87793 (1968)
237) Airapetyants, A. V., Vlasova, R. M., Geiderikh, M. A., Davydov, B. E.: Izv. Akad. Nauk SSSR., Ser. Khim. **1964**, 1328; C.A. **61**, 13440 (1964)
238) Dinh-Suang-Dinh: Dissertation for the Degree of Candidate of Sciences, Institute of Petrochemical Synthesis, USSR Academy of Sciences, Moscow 1971
239) Dinh-Suang-Dinh, Geiderikh, M. A., Davydov, B. E.: Izv. Akad. Nauk SSSR, Ser. Khim. **1970**, 2033; C.A. **73**, 131456 (1970)
240) Kubasova, N. A., Dinh-Suang-Dinh, Geiderikh, M. A., Shishkina, M. V.: Vysokomol. Soedin., Ser. A, **13**, 162 (1971); C.A. **74**, 76912 (1971)
241) Geiderikh, M. A., Dinh-Suang-Dinh, Davydov, B. E., Karpacheva, G. P.: Vysokomol. Soedin., Ser. A, **15**, 1239 (1973); C.A. **79**, 79582 (1973)
242) Silin, E. A., Motorykina, V. P., Shmit, J. K., Geiderikh, M. A., Davydov, B. E., Krentsel, B. A.: Elektrokhimiya **2**, 117 (1966); C.A. **64**, 14347 (1966)
243) Silin, E. A., Motorykina, V. P., Geiderikh, M. A., Davydov, B. E., Krentsel, B. A.: Zh. Fiz. Khim. **41**, 309 (1967)

244) Kubasova, N. A., Shishkina, M. V., Kusakov, M. M.: Prikladnaya spektroskopiya (applied spectroscopy). Vol. 2. Moscow: Nauka, 1969, p. 105
245) Brandrup, J., Kirby, J. R., Peebles, L. H.: Macromolecules 1, 59, 64 (1968)
246) Silin, E. A., Ekmane, A. Ya., Khutareva, G. V., Davydov, B. E.: Vysokomol. Soedin., Ser. A, 10, 1786 (1968); C.A. 69, 77908 (1968)
247) Pen'kovskii, V. V., Kruglyak, Yu. A.: Zh. Strukt. Khim. 10, 459 (1969); C.A. 71, 70948 (1969)
248) Samedova, T. G., Karpacheva, G. P., Davydov, B. E.: Zh. Prikl. Spektrosk. 13, 431 (1970); C.A. 74, 13532 (1971)
249) Davydov, B. E., Karpacheva, G. P., Samedova, T. G., Yandarova, M. M.: Eur. Polym. J. 7, 1569 (1971)
250) Davydov, B. E., Zakharian, R. Z., Karpacheva, G. P., Krentsel, B. A., Lapitskii, G. A., Khutareva, G. V.: Dokl. Akad. Nauk SSSR 160, 650 (1965); C.A. 62, 13258 (1965)
251) Davidov, B. E., Krentsel, B. A., Radzhabli, N. A., Aliev, A. D.: Vysokomol. Soedin., Ser. B, 12, 326 (1970). C.A. 73, 46151 (1970)
252) Karpacheva, G. P., Davydov, B. E., Aliev, A. D.: Vysokomol. Soedin., Ser. B, 13, 729 (1971); C.A. 76, 60246 (1972)
253) Samedova, T. G.: Dissertation for the Degree of Candidate of Sciences, Institute of Petrochemical Synthesis, USSR Academy of Sciences, Moscow 1973
254) Benderskii, V. A., Stunzkas, P. A.: Vysokomol. Soedin. 6, 1104, (1964); C.A. 61, 8430 (1964)
255) Samedova, T. G., Karpacheva, G. P., Davydov, B. E.: Izv. Akad. Nauk SSSR., Ser. Fiz. 36, 1129 (1972); C.A. 77, 75622 (1972)
256) Samedova, T. G., Karpacheva, G. P., Davydov, B. E.: Eur. Polym. J. 8, 599 (1972)
257) Samedova, T. G., Gavrilenko, J. F., Kapacheva, G. P., Davydov, B. E., Stefanovskaya, N. N.: Izv. Akad. Nauk SSSR., Ser. Khim. 1970, 682
258) Davydov, B. E., Korshak, Yu. V., Krentsel, B. A.: Dokl. Akad. Nauk SSSR 157, 611 (1964); C.A. 61, 10794 (1964)
259) Gugeshashvly, M. J., Davydov, B. E., Korshak, Yu. V., Rozenshtein, L. D.: Izv. Akad. Nauk SSSR, Ser. Khim., 1964, 1703
260) Grigorovskaya, W. A., Sel'skaya, O. G., Astrachanskaya, N. J., Berlin, A. A.: Plaste Kautsch. 21, 897 (1974)
261) Yandarova, M. L.: Dissertation for the Degree of Candidate of Sciences, Institute of Petrochemical Synthesis, USSR Academy of Sciences, Moscow 1970
262) Yandarova, M. L., Geiderikh, M. A., Krentsel, B. A.: Izv. Akad. Nauk SSSR, Ser. Khim. 1970, 78; C.A. 72, 111851 (1970)
263) Khutareva, G. V., Yandarova, M. L., Shishkina, M. V.: Vysokomol. Soedin., Ser. B, 12, 515 (1970); C.A. 73, 88367 (1970)
264) Khutareva, G. V., Orlova, O. V., Davydov, B. E., Boguslavskii, L. J.: Vysokomol. Soedin., Ser. A, 9, 772 (1967); C.A. 67, 64790 (1967)
265) Brandrup, J.: Macromolecules 1, 72 (1968)
266) Davydov, B. E., Samedova, T. G., Karpacheva, G. P.: Preprints XXIII International Symposium on Macromolecules, Madrid 1974, VI, p. 42
267) Karpacheva, G. P.: Dissertation for the Degree of Candidate of Sciences, Institute of Petrochemical Synthesis, USSR Academy of Sciences, Moscow 1966
268) Krasnovskii, A. A., Brin, G. P.: Dokl. Akad. Nauk SSSR 53, 447 (1946)
269) Vol'kenshtein, F. F.: Elektronic theory of catalysis on semiconductors. Moscow: Fismatgiz 1960
270) Topchiev, A. V., Geiderikh, M. A., Davydov, B. E., Kargin, V. A., Krentsel, B. A., Kustanovich, J. V., Polak, L. S.: Dokl. Akad. Nauk SSSR 128, 312 (1959)
271) Dokukina, E. S., Roginskii, S. Z., Sakharov, M. M., Topchiev, A. V., Geiderikh, M. A., Davydov, B. E., Krentsel, B. A.: Dokl. Akad. Nauk SSSR 137, 893 (1961); C.A. 56, 2916 (1962)
272) Berlin, A. A., Blumenfeld, L. A., Semenov, N. N.: Izv. Akad. Nauk SSSR, Otd. Khim. Nauk. 1959, 1689

273) Keier, N. P., Boreskov, G. K., Rode, V. V., Teren'ev, A. P., Rukhadze, E. G.: Kinetika i Kataliz **2**, 509 (1961); C.A. **58**, 5082 (1962)
274) Gallard, J., Traynard, Ph.: C. R. Hebd. Seances Akad. Sci. **254**, 3529 (1962)
275) Manassen, J., Wallach, J.: J. Am. Chem. Soc. **87**, 2671 (1965)
276) Dokukina, E. S., Golovina, O. A., Sakharov, M. M., Aseeva, P. M.: Kinetika i Kataliz **7**, 660 (1966); C.A. **65**, 17755 (1966)
277) Teren'ev, A. P., Rode, V. V., Rukhadze, E. G.: Vysokomol. Soedin. **2**, 1557 (1960); C.A. **55**, 19303 (1961)
278) Hock, H., Kropf, H.: J. Pract. Chem. **9**, 173 (1959)
279) Kropf, H., Justus Liebigs Ann. Chem. **637**, 93, 111 (1960)
280) Klier, N. P., Boreskov, G. K., Rubtsova, L. F., Rukhadze, E. G.: Kinetika, i Kataliz **3**, 680 (1962); C.A. **58**, 7412 (1963) Boreskov, G. K., Keier, N. P., Rubtsova, L. F., Rukhadze, E. G.: Dokl. Akad. Nauk, SSSR **144**, 1069 (1962); C.A. **57**, 13959 (1962)
281) Davydova, J. R., Kiperman, S. L., Slinkin, A. A., Dulov, A. A.: Izv. Akad. Nauk SSSR., Ser. Khim., **1964**, 1591; C.A. **62**, 7155 (1965); Kiperman, S. L., Davydova, J. R.: Kinetika i Kataliz **2**, 762 (1961); C.A. **56**, 12352 (1962)
282) Gallard, J., Laederich, T., Salle, R., Traynard, Ph.: Bull. Soc. Chim. Fr., **1963**, 2204, 2209
283) Paushkin, Ja. M., Burova, L. M., Amelekhina, L. N., Alekseev, S. N.: Dokl. Akad. Nauk SSSR **180**, 367 (1968); C.A. **69**, 43328 (1968); Paushkin, Ja. M., Burova, L. M., Lunin, A. F., Vishnyakova, T. P., Gornshtein, A. D., Sladkov, A. M., Neposhatykh, V. P., Korshak, V. V.: Izv. Akad. Nauk SSSR, Ser. Khim., **1969**, 1073; C.A. **71**, 60435 (1969)
284) Terpugova, M. P., Mazur, V. G., Kotlyarevskii, J. L.: Izv. Akad. Nauk SSSR., Ser. Khim., **1971**, 2097; C.A. **76**, 3149 (1972)
285) Roginskii, Z.: Zh., Vses. Khim. Ova. **5**, 482 (1969)
286) Vogel, H., Marvel, C. S.: J. Polym. Sci. **50**, 511 (1961)
287) Dawans, F.: Proceedings of the Conference on High-Molecular Compounds, Paris 1965
288) Kovaleva, V. P., Kukina, E. D., Kabanov, V. A., Kargin, V. A.: Vysokomol. Soedin. **6**, 1676 (1964); C.A. **61**, 16170 (1964)
289) Berlin, A. A., Vonsyatsky, V. A., Lyubchenko, L. S.: Izv. Akad. Nauk SSSR, Otd. Khim. Nauk, **1962**, 1312; C.A. **58**, 3021 (1963); Berlin, A. A.: Izv. Akad. Nauk SSSR, Otd. Khim. Nauk, **1965**, 59; C.A. **62**, 16028 (1965)

Received June 14, 1976; May 9, 1977
W. Kern (editor)

The Behaviour of Furan Derivatives in Polymerization Reactions*

Alessandro Gandini

Division of Chemistry, National Research Council of Canada, Ottawa, K1A 0r6 Canada

Table of Contents

I. Introduction 49
II. Polycondensation 50
   A. Linear Polyesters 50
   B. The Resinification of 2-Furfuryl Alcohol 52
   C. The Thermal Resinification of 2-Furaldehyde 54
III. Polyaddition 56
   A. Polymerization through the Ring 57
      1. Furan and 2-Alkylfurans 57
         a) Free-Radical Systems 57
         b) Anionic Systems 57
         c) Cationic Systems 58
         d) γ-Ray Irradiation 61
         e) Stereospecific Systems 61
      2. Copolymerization of Furan and Homologues with Maleic Anhydride 62
      3. Benzofuran and Naphthofurans 63
      4. Dihydrofurans 65
      5. The Photopolymerization of 2-Furaldehyde 66
      6. Miscellaneous Reactions 68
         a) The Paterno-Büchi Reaction with Furans 68
         b) The Aromatization of bis-Furan Adducts 68
         c) Poly(2,5-difurylketone) 68
   B. Polymerization through a Function External to the Ring . . . 69
      1. 2-Alkenylfurans 69
         a) Free-Radical Systems 69
         b) Anionic Systems 71

---

* Issued as NRCC No. 16042

	c) Cationic Systems	72
	d) γ-Ray Irradiation	75
	e) Stereospecific Systems	75
	2. Other Olefins Containing the Furan Ring	76
	3. Vinyl Esters of the Furan Series	76
	4. Furfuryl Acrylate and Furfuryl Methacrylate	78
	5. Vinyl Ethers and Vinyl Carbonyl Compounds of the Furan Series	79
	a) Vinyl Ethers	79
	b) Unsaturated Aldehydes and Ketones	80
	6. Polymerization of 2-Furaldehyde and Homologues through the Carbonyl Bond	81
IV.	The Transformation of Poly(vinylalcohol) into Furan Polymers	85
V.	The Retarding and Inhibiting Role of Furan Derivatives in Radical Polymerization	85
	A. Strong Generalized Inhibition	86
	B. Selective Retardation	86
	C. Autoinhibition	88
VI.	Stability and Resinification of Furan Derivatives	89
VII.	Conclusions and Acknowledgments	92
VIII.	References	93

I. Introduction

The fact that many vegetable residues such as oat hulls, corn husks and sugar-cane bagasse, etc., are raw materials from which 2-furaldehyde can be obtained in good yields makes this compound and its derivatives easily accessible in large quantities. Considering moreover that these raw materials are naturally renewed, the synthesis of furan derivatives becomes an attractive alternative to pretroleum-based technologies. One of the various uses of these compounds which has been exploited to a certain degree, but which certainly has wider potential, is the preparation of polymeric products.

Five-membered heteroaromatic ring compounds have been extensively studied in terms of their relative reactivity and specific physicochemical properties. In particular, it is generally accepted that the degree of aromaticity follows the order

thiophene > pyrrole > selenophene > furan[1],

while on the other hand their dienic character follows approximately the reverse order, as shown by studies on the Diels-Alder and other addition reactions[2].

Furan, therefore, shares both aromatic and dienic properties. The predominance of one of these two effects among its derivatives depends essentially upon the nature and position of the substituents and upon the specificity of the reagents, catalysts and media used in a given situation. Similar considerations will apply to the degree of participation of the ring when furan derivatives are involved in reactions which are sought to occur only on the substituent(s), *i.e.* in some cases the progress of the reaction is clean and selective and the ring remains essentially untouched; in others, the interference of side reactions involving ring addition, substitution, oxidation or cleavage can be severe.

The complex and rather unique chemistry of furan derivatives[3] and the instability of many of these compounds including their proneness to resinify spontaneously or in mild media (see Chapter VI) are the broad consequences of the aromatic-dienic duality displayed by the heterocycle. Chain or step polymerization reactions of monomers possessing the furan ring will inevitably reflect this duality in their kinetic and mechanistic behaviour and ultimately in the structure and molecular weights of the products obtained. Moreover, the stability of these polymers towards atmospheric degradation (oxidation, photolysis, etc.) will also depend upon the "state" of the ring in the macromolecule, *i.e.* on the position and nature of its substituents, on the possible presence of modified rings (*e.g.* dihydrofuranic structures), etc.

The present review deals with polymerization systems where the furan ring is present in the monomer(s) either as *the* reactive entity or as a side group to the function responsible for the growth. It covers, therefore, a wide variety of situations, many of them not yet fully understood. In fact, as will become apparent later, there is still a great deal of controversy about the interpretation of experimental results obtained with most of these systems and sometimes even disagreement among authors as to the data obtained. Particular emphasis has therefore been placed in this review on a critical reinterpretation of previous work in view of the recent experience gathered by the author and his colleagues during several years of research on these topics at the Polymer Laboratory, *Centro Nacional de Investigaciones Cientificas in Havana, Cuba.*

A vast amount of technological reports dealing with furanic resins of various composition has appeared in the literature. Examples of these materials are 2-furaldehyde resins with urea, formaldehyde, phenols, etc., modified by appropriate binders and fillers. Usually, the products of these resinifications are cross-linked materials which prove useful in specific applications such as the building industry, new moulding technologies etc., depending on their composition and on the nature of the additives employed. This field is mostly covered by patents. The chemical processes which take place in the course of these complex polymerization reactions have not been studied and are not likely to be elucidated in the near future. These processes and materials are not reviewed here. Attention has instead been focused on polymerizations where conditions have or can be controlled, conversions followed, structures determined, products analysed and/or molecular weights measured. Thus, the present review deals with systems that have provided some information which should be of interest to the physical-organic polymer chemist engaged in kinetic and mechanistic studies.

The field has been subdivided according to the classical polycondensation-polyaddition structure. Polyaddition products have in turn been classified in two broad categories,

a) those which bear the furan ring (or some cyclic structure derived from it) in the chain's backbone and

b) those in which the ring is pendant to the chain.

This second classification is not rigorous since often the polymer structure is not defined by only one type of repeat unit and the furan ring is encountered both in the backbone and as a side group. However, it is felt that the practical convenience of this classification outweight its minor inconsistencies.

Although an extensive bibliographic research was carried out in order to make this review as comprehensive as possible, only those references are given which are pertinent to the aim stated above. Therefore, some publications of a merely descriptive nature (particularly before 1960) have been omitted, but will be found quoted in the relevant papers which were selected for discussion.

Throughout this paper Fu stands for a 2-furylic group, [furan ring structure].

II. Polycondensation

All polymers described in this chapter bear the furan ring in the chain's backbone, and are the result of step growth reactions which involve the elimination of a condensation product.

A. Linear Polyesters and Polyamides

The use of furan bifunctional monomers for the preparation of linear polyesters and polyamides has been reported by several authors in the last two decades and two recent bibliographies are available[4,5]. The vast majority of these studies describe the use of 2,5-disubstituted monomers. The classical starting material for these preparations has been 2,5-furan dicarboxylic acid, an easily synthesised and stable compound which has been employed directly or as its dichloride, diesters, or dihyrazide. 2,5-*bis* (hydroxymethyl) furan has also been used as a monomer as well as unsymmetrically substituted derivatives such as 5-hydroxymethyl-2-furoic acid and 5-aminomethyl-2-furoic acid. Recently, 3,4-disubstituted furans have also been reported as monomers in the preparation of polyamides[6].

In general, the research in this field has been qualitative and aimed at obtaining products with properties comparable to those of well-known aromatic counterparts prepared with monomers such as the phthalic acids, Bisphenol A, etc. Virtually all the standard techniques of polycondensation, including bulk, solution and interfacial mixing have been tried with a wide variety of catalysts in an effort to optimize the quality of the polymers. There is a great deal of disagreement among data obtained by different authors dealing with a given system, the more recent studies being in conflict with older ones[4,5]. This situation makes it difficult to establish clear-cut conclusions on the role played by the furan ring in these polymerizations. Nevertheless, some recurring observations make it possible to establish a general pattern of behaviour. Thus, the presence of the furan ring in the monomer(s) can lead to the following problems:

— Some of the typical conditions of polycondensations used for aliphatic and aromatic monomers are not suitable for furan derivatives, *e.g.*, the melt polycondensation of 2,5-furan dicarboxylic acid chloride with 2,5-*bis*(hydroxymethyl) furan at about 80 °C only yields a black insoluble product[5]. The hydrochloric acid liberated in the reaction is clearly responsible for the charring of the furanic diol which like its simpler homologue furfuryl alcohol, resinifies rapidly in acidic media (see below).

— It is a rather difficult experimental task to achieve reasonably high molecular weights with furanic monomers[5]. The causes of this difficulty are probably to be found in the occurrence of side reactions which involve the functional groups of the growing molecule. No specific study of the origin of this problem has been undertaken, but empirical modification of parameters and reaction conditions had led sometimes to higher DP's.

— The products of polycondensations involving furanic monomers are often brown, even after repeated reprecipitation, *i.e.* their colour is intrinsic to the macromolecular structure which has already suffered some degradation during the synthesis. Connected with this darkening is a lack of crystallinity as revealed by X-ray analysis on drawn samples and by thermograms which show no definite transition[5]. Occasional branching caused by side reactions introducing new functionalities and/or partial attack of the furan ring with formation of chromophores could account for this poor behaviour.

— Most of these furan polycondensates are more sensitive to thermal and oxidative degradation than their benzene counterparts. Particularly affected are the polyesters obtained from 2,5-*bis*(hydroxymethyl) furan indicating that one of the vulnerable groups must be the -Fu–CH_2–O-, and not the -Fu–CO–O-, since polycondensates obtained from 2,5-dicarboxylic acid are more stable, as expected from the stability of the simple furoates. The methylene group next to the ring must be the cause of this lability judging from the instability of compounds such as furfuryl alcohol, furfuryl chloride, furfuryl nitrile, etc. (see Chapter VI).

Notwithstanding these drawbacks, some interesting polycondensates have been prepared from furanic monomers. These have reasonably high molecular weights, crystallize satisfactorily and can be drawn to give fibers: hexamethylene diamine and 2,5 furandicarboxylic acid give one such polyamide[4] and ethylene glycol with 2,5-furandicarboxylic acid one such polyester[5]. These and similar polymers give good thermograms with clear first-order transitions[5].

There is clear need for a more physicochemical approach to these systems; as will be pointed out appropriately, this need is evident in many other areas of the field covered by the present review, where interesting aspects are merely touched upon without any further effort to gain a better understanding of the rather intricate problems arising from that first approach. Most of the older work on the polycondensation of furan derivatives was qualitative and superficial, with rare exceptions (see Ref.[5] for a good survey), and only recently have more systematic studies been carried out[4, 5, 7]. The emphasis, however, must be moved onto kinetic and mechanistic investigations and onto a more detailed look at the polymers' structure, with particular attention to the effect of side reactions. This would lead to more reliable conclusions and ultimately to solution of the problems related to the often inferior qualities of the products. Also, such work would be of considerable interest to the general field of furan chemistry.

B. The Resinification of 2-Furfuryl Alcohol

One of the most important "furan resins" from an industrial standpoint is undoubtedly that obtained from 2-furfuryl alcohol. The final cross-linked product displays outstanding chemical, thermal and mechanical properties[8].

The chemistry of the self-condensation processes promoted by heat, acidic catalysts and alumina has been investigated in various laboratories and has proven rather complex, particularly after the early stages of the polymerization. Some basic evidence about the reactions involved in this resinification was gathered a few decades ago and is clearly summarized by Dunlop and Peters in their classical monograph on furans[3]. More recently, this topic has been revived by studies dealing with the unequivocal identification of a number of intermediates by spectroscopic and chromatographic techniques[9–11], the kinetics of the process[12, 13], the structure of the polymers[14] and the nature of the oxygen- or acid-catalysed cross-linking of the initial resin[15]. The original scheme proposed by Dunlop and Peters has not been substantially challenged by these developments, but a better and more solid understanding of the general mechanism has certainly been achieved. This is briefly given below.

The first step involves a dimerization which can occur by head-to-tail or head-to-head condensation, the former being favoured:

The electrophylic attack at C-5 predominates over the C-3 and C-4 possibilities to such an extent that no intermediates containing 3- or 4- substituted rings have ever been isolated. This is in agreement with a large body of experimental evidence[16] and theoretical considerations[17]. For this reason 2-furfuryl alcohol must be considered a bifunctional monomer in the initial stage and its "normal" reactions will give linear chains. The two dimers produced in the first condensation step will therefore grow through successive condensations with 2-furfuryl alcohol to yield a mixture of linear oligomers containing essentially two repeat units, *viz.*

$$\text{furan}-CH_2-O-CH_2-$$

and

$$\text{furan}-CH_2-,$$

the latter predominating. Many of these oligomers (with up to five furan rings) have been isolated from reaction mixtures prepared under controlled conditions[9-11]. The end groups of these oligomers were $-CH_2OH$, unsubstituted furan rings and methyl groups. The presence of 2,2'-difurylmethane and of higher homologues indicates that the formaldehyde detected as a gaseous product of the reaction is released by the terminal alcoholic groups and/or the internal ether bridges. The methyl end groups are more difficult to explain and Barr and Walton, the only authors who isolated such oligomers avoided commenting on this[10]. It is conceivable that if the furfurylenium ion *1* is formed as an intermediate, it could react with the catalyst's anion, A^-, and ultimately give compound *2*[18], which would participate in the polycondensation thus giving rise to some oligomers with a 2-methyl end group:

$$[\text{furan}=CH_2]^+ + A^- \longrightarrow \overset{H}{\underset{A}{\text{furan}}}=CH_2 \longrightarrow A-\text{furan}-CH_3$$

1 *2*

Kinetic observations of the homogeneous part of the reaction in water[12, 13] do not provide any substantially new element to the knowledge of this system. The obvious observations that the rate of resinification increases with increasing temperature and decreasing pH of the mixture only provide technically useful correlation parameters and the zero-order of reactions carried out to small conversion of 2-furfuryl alcohol[13] does not indicate anything except an elementary kinetic approximation (the use of colour build-up as a criterion for the extent of alcohol consumed is also questionable since no firm relationship has ever been established between these two quantities).

The resinification induced by γ-alumina[11] seems to proceed by a somewhat different mechanism, probably because of the higher temperatures involved. Side reactions are more prominent from the beginning and it has been suggested, but not proved, that the C-3 and C-4 positions of the ring are vulnerable under these conditions to substitution reactions. A new compound, *viz.* 4-(2-furfuryl)-2-pentenoic acid-γ-lactone, which is an isomer of the condensed dimers, was identified among the products of these reactions[11]. A mechanism for its formation was proposed[11b], but levulinic acid, the precursor envisaged, was not detected in the system. This acid is known to form by hydrolytic ring cleavage of 2-furfuryl alcohol, but has never been isolated in resinification studies.

The problems really begin when one tries to rationalize the evidence which shows that the oligomers formed in the first phase are polymerized further to form a cross linked resin. This reaction is catalysed by acids and accelerated by heat, oxygen and the addition of various compounds, e.g. formaldehyde. As argued above, if only oligomers with the structures proposed were present in the first-stage product, i.e. compounds with functionality two, one would only expect a further linear growth in the second stage with attainment of high molecular weights. Likewise, none of the structures proposed, and indeed identified, justify the progressive darkening of the reaction mixture, as no strong chromophore or high degree of conjugation are present. Conley and Metil[14] have argued that the curing of the resin in a nitrogen atmosphere only brings about the further condensation of the oligomer to linear polymers. As for the cross-linking of the resin, they could not offer any definite explanation. In a later paper[15] these authors studied the oxidative curing of the resin and proposed a mechanism involving the formation of carbonyl groups from methylenic bridges and ultimately chain scission to give carboxylic end groups. None of the reaction schemes proposed could explain cross-linking or colour build-up, however, as neither third functionalities in a given chain nor the formation of any chromophores was postulated. Nakamura and Saito[19] examined the curing of a model compound, 2-hydroxymethyl-5-(5'-furfuryl) furfuryl furan, i.e. one of the oligomers formed in the initial stage of the reaction, and discussed the formation of cleaved compounds produced by the action of oxygen and acids on the furan ring and their role in promoting cross-linking. Polyconjugated structures and possibilities of branching were clearly envisaged and although not all of their conclusions were substantiated by the experimental results, there was a good correlation between the phenomenology of the curing (reticulation and darkening of the resin) and the arguments put forward to explain it. Wewerka[11b] has also discussed the mechanism of curing and has proposed the participation of the C-3 and C-4 ring positions in the acid-catalyzed process. The curing with γ-alumina was thought to proceed via occasional ring cleavage at temperatures above 200 °C.

It can be concluded that the polycondensation of 2-furfuryl alcohol is essentially a bifunctional, linear aggregation process, accompanied by the occurrence of side reactions accelerated by acids, oxygen and heat, which introduce occasional branching points and polyconjugated sequences in the chains. These side reactions give rise to ring cleavage and/or oxidation and structural rearrangements at some repeating units, and ultimately to the cross-linking and blackening of the products.

C. The Thermal Resinification of 2-Furaldehyde

The resinification of 2-furaldehyde promoted by acidic substances or by heat has been known to chemists since the end of last century, and attempts to explain the mechanism leading to the formation of black, insoluble resins have been published since 1919[20]. Dunlop and Peters[3, 21], Illari[22] and Nakamura and Saito[23], among others, followed the original work of Marcusson. The major drawback in establishing a mechanism in all these studies has been the lack of identified intermediates. Also, there has been a tendency to group together indiscriminately resins formed by heat, aqueous acids and by acidic catalysts under anhydrous conditions. There are at present good reasons to believe that although all these resins "look alike", the processes involved in their formation could be different[24, 25]. Thus, it is obviously not the same to treat 2-furaldehyde with a dilute acid in the presence of air or to carry out its resinification with, say, $BF_3 \cdot Et_2O$ in an inert atmosphere, since the hydrolytic cleavage possible in the first situation will not take place in the second and similarly the intervention of atmospheric oxygen can change appreciably the nature of the reactive intermediates. Recent work on the resinification of 2-furaldehyde[24, 25] has shed some new light on these problems, particularly with respect to the thermal reaction at 100–250 °C in the absence of air. This specific topic will be discussed presently, while other aspects related to resinification by acids etc., will be dealt with in Chapter VI.

Two intermediate products were isolated in the study of the thermal resinification of 2-furaldehyde[24, 25]. They were characterized by spectroscopic and other standard techniques and their structures are given below:

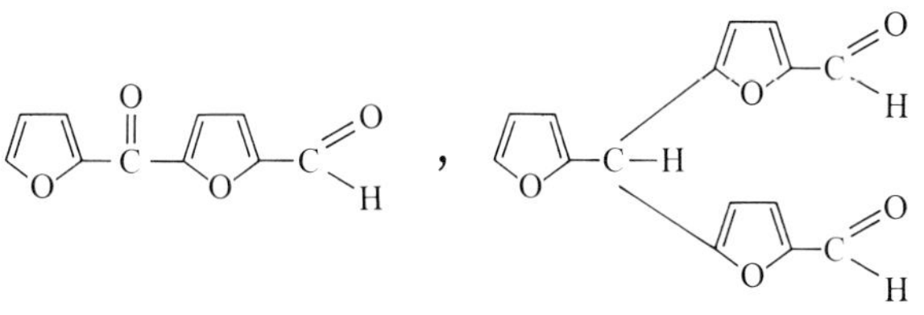

3   *4*

The condensation product *4* was relatively more abundant. Low-molecular-weight soluble resins were also studied for the first time and their structural characteristics indicated that they were branched oligomeric polycondensates containing both precursors *3* and *4*. The resin at advanced stages of condensation was black even before total cross-linking and EPR spectra of these powders gave intense signals of unpaired electrons probably delocalized over several conjugated units of the branched chains. These observations explain the intense colour. The combined study of the precursors and the resins at various stages of condensation, together with the kinetics of resin and colour accumulation and the quantitative determination of the water produced allowed the formulation of an overall mechanism based upon the following points:

– an equilibrium initiation reaction to give compound *4* and water;
– reaction of *4* with *3* and with more 2-furaldehyde to give oligomers and more water;
– removal of the tertiary hydrogen of *4* (or of equivalent oligomers) to give unpaired electrons distributed over conjugated structures (colour);
– further growth and precipitation of the cross-linked resin;
– the achievement of a stationary state between the rate of initiation (initial condensation) and that of termination (precipitation of gel) with the colour inten-

sity and the concentration of intermediate oligomers remaining practically constant. The structure of the final product must be close to that represented below:

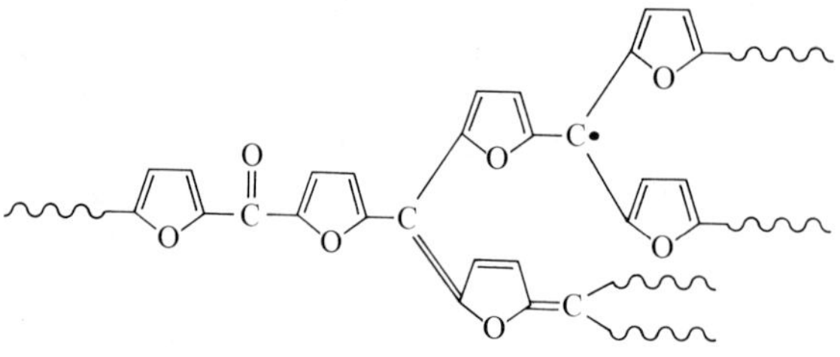

5

The presence of stable free radicals in the resin was further suggested by the strong inhibiting effect of traces of this product on the thermal polymerization of styrene.

The catalytic effect of various surfaces was also investigated and showed that electron-deficient sites were responsible for promoting the condensation. Basic substances and small amounts of water were found to considerably reduce the rate of resinification.

In conclusion, the self-condensation of 2-furaldehyde promoted by heat occurs with the formation of di- and trifurylic intermediates. The functionality of the growing chain increases after each oligomerization step until gelation and precipitation of the resin occurs. Thus, the process is non-linear from the onset since the condensation product *4* possesses three sites for further attack, namely the free C-5 position and the two formyl groups. It is interestering to note that while the polycondensation of 2-furfuryl alcohol is essentially linear and cross-linking is due to side reactions, the thermal resinification of 2-furaldehyde is intrinsically non-linear and gel formation occurs at earlier conversions.

III. Polyaddition

The furan ring can be made to polymerize through one or both of its double bonds and the polymers obtained will therefore have dihydro- and tetrahydrofuran rings in their backbone. This situation occurs when furan, the alkylfurans, benzofuran and some dihydrofurans are treated with suitable initiators and is discussed in the first section of this chapter.

If on the other hand the polymerization of a furan derivative takes place through a substituent containing an adequate functionality, such as C=C or C=O, the furan ring should in principle conserve its structure and the polymers obtained will bear it as a side group. It has been found, however, that in some of these systems the normal propagation is accompanied by other reactions which involve participation of the ring and which therefore alter the normal structure of the macromolecule. The second section of this chapter deals with monomers, such as 2-vinylfuran and 2-furaldehyde, which exhibit this general behaviour.

A. Polymerization through the Ring

1. Furan and 2-Alkylfurans

a) Free-Radical Systems. The reaction of furan and some of its derivatives with various free radicals has been studied (see [26] and references therein). Addition and/or substitution products are obtained and the relative importance of these two modes of interaction depends largely on the primary radicals used. No polymerization of the furan ring has ever been reported and the reasons for this failure must be connected with the stability of the radical formed in the initial reaction[26]:

6

The allylic-type furylic radical 6 is resonance stabilized to such a degree that its reactivity in promoting propagation by adding onto another furan ring is minimal. The fate of these radicals will simply be to couple with another radical present in the reaction medium (primary or secondary) or to disproportionate to regenerate the furan character of the ring[26].

The broader implications of this behaviour will be discussed in Chapter V.

b) Anionic Systems. The reaction of furan and 5-methylfuran with alkali metals only takes place if Na-K alloy is used. Metallation takes place at C-2 as indicated by the isolation of 2-furoic acid after carbonation[27]. The use of butyl lithium gives considerably higher yields and 2-furyl and 5-methyl-2-furyl lithium have been used successfully in a number of synthetic preparations[28, 29]. Normant and Angelo[30] and Goutière and Golé[31] have observed the formation of oligomers of furan and 2-ethylfuran when these are treated with Na-K alloy or (more efficiently) with sodium naphthalene. The yields are very low even when an excess of sodium naphthalene is used and the reaction allowed to proceed for ten days at room temperature. The structure of these oligomers was not investigated. This weak reaction involving the furan ring and sodium naphthalene was confirmed by the observation that a solution of poly(2-vinyl-furan) treated with this compound produced about 10% of cross-linking in a few hours[31]. It must be pointed out that these reactions were carried out with very high concentrations of sodium naphthalene and should therefore be regarded as very inefficient. It can be concluded that the furan ring is insensitive to anionic polymerization, *i.e.* that the furyl anion does not induce an efficient chain propagation of its neutral precursors.

A recent study of the interaction of sodium with furan in an argon matrix at 4 °K[32] has shown that the irradiation of the matrix with visible light ($\lambda = 550$ nm) induces the reaction

$$Na + FuH \xrightarrow{h\nu} Na^+ + FuH^-$$

The ESR spectrum of the furan radical anion indicates that the Cem-0 bond is ruptured in the electron transfer process whereby the oxygen atom acquires the negative charge and the C-2 end of the open ring possesses a free radical character:

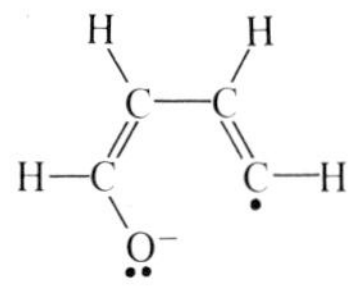

7

Of course in solution such an intermediate, if formed at all, would be very short-lived and could be the species responsible for the "side reaction" which produces some oligomerization of the furan ring. However, any interpretation of the mechanism of oligomerization must be regarded at this point as highly speculative since there is an obvious need for more experimental evidence, particularly concerning the structure of the products.

c) Cationic Systems. Acidic catalysts are known to promote the polymerization of furan and alkylfurans. Several studies of the mechanism of these processes, and of the structure and properties of the products have been published. Both Lewis and Brønsted acids are effective initiators but fairly high concentrations are needed to achieve reasonable yields. The polymerization of furan by trichloroacetic acid (conc. 2M, room temperature, 170 h, no solvent: 20% yield) was investigated by Wassermann and co-workers[33] who placed particular emphasis on the polyconjugated character of the products, their high proton affinity and their complex structure containing several possible repeating units and trichloracetate groups. These authors also pointed out that branching must have taken place during the polymerization. Kresta and Livingston[34] continued this work and arrived at the formulation of both a mechanism and the most probable and frequent structural units in polyfuran. They also reported that cross-linking in the growing chains depends on the acidity of the catalyst employed: thus, while trichloroacetic acid yields soluble polymers, perchloric acid gives totally insoluble products and trifluoroacetic acid produces a mixture of the two. In another paper[35] these authors described the polymerization of gaseous furan on the liquid surface of concentrated perchloric acid. The mechanism proposed[34] to account for the observations obtained in the two laboratories[33−35] has an initiation step involving the equilibrium protonation of furan at C-2 with formation of the unstable furyl carbenium ion pair.

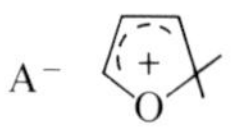

8

The propagation involves the attack of this species onto the C-2 position of another furan molecule leading to the growth of a chain with the following structure:

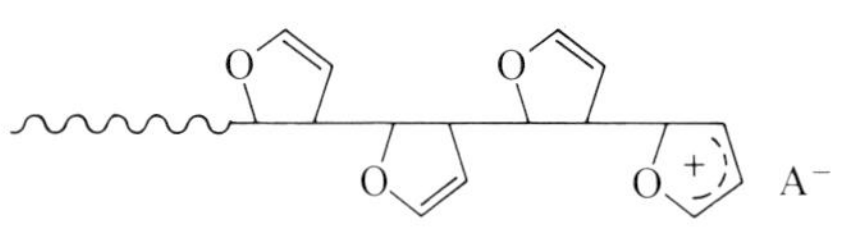

9

However, the 2 3-dihydrofuran units in these chains are readily susceptible to further attack by the acid or a growing species at the 4,5-unsaturation (see Section III-4) and this produces branching and the formation of tetrahydrofuran rings:

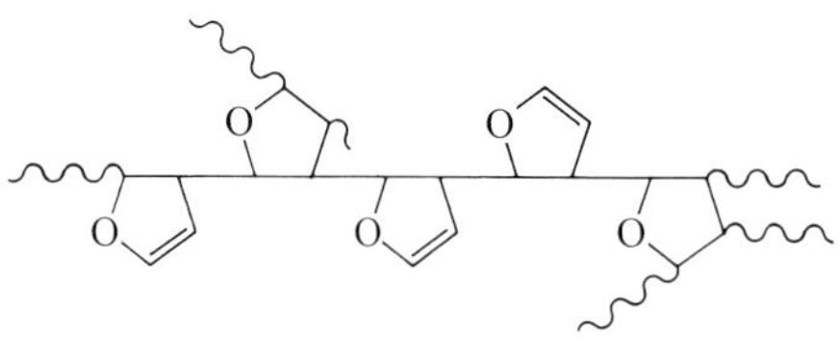

10

Finally, two termination steps are considered: a spontaneous termination with acid expulsion and restoration of a terminal 2,3-dihydro- or furan ring; and a recombination of the growing ion pair to give a tetrahydro- or a 2,3-dihydrofuran ester.

All structures envisaged in the above mechanism were detected by spectroscopic techniques and other analytical tools[33, 34].

The approximate relative abundance of each one indicated that normal propagation and chain branching reactions are competitive, as shown by the ratio of 2,3-dihydro- and tetrahydrofuran rings in the polymers. Moreover, the spectroscopic observation that the furan rings are still an important proportion of all rings present[34] also indicates a high degree of branching, *i.e.* a large number of chain ends *per* polymer molecule, considering that the ester end groups were only 5–10% of all terminal structure and the DP's ranged from 50 to 300 in both sets of experiments[33, 34].

The colour of these polymers, their electronic spectra and their high proton affinity suggested the presence of conjugated structures. The acid-catalyzed isomerization of 2,3-dihydrofuran rings to 2,5-configurations and some ring cleavage with formation of carbonyl bonds were advocated to explain the formation of such structures[34]:

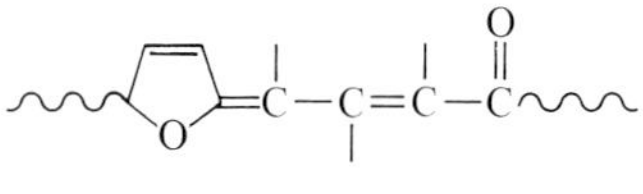

11

The spontaneous polymerization of furan adsorbed on carbon black with or without $SnCl_4$ vapours[35] has been explained by a similar cationic mechanism. Also, the polymerization of gaseous furan on liquid acidic surfaces[35] has the same origin, but in these systems the polymers suffer an acid-catalyzed hydrolysis of their tetrahydrofuran rings which produces a considerable proportion of hydroxyl and carbonyl groups.

2-Methylfuran also polymerizes with acidic catalysts and in fact it has been shown to be more prone to undergo this process than furan[36]. Korshak, Sultanov and Abduvaliev[37] were the first to propose a structure for the cationically prepared poly(2-methylfuran), but their poly(2,5-dihydro-2-methylfuran) model was soon challenged[38] and thereafter they also revised it[39]. The new structure proposed on the basis of evidence from IR spectra and ozone-H_2O_2 degradation was poly(2,3-dihydro-2-methylfuran):

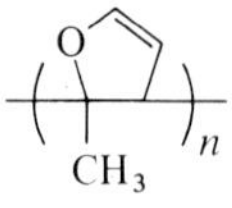

12

In all these investigations Lewis acids were used as initiators at temperatures between −30 and 60 °C. The arguments used to substantiate the validity of structure *12* are unconvincing to this reviewer particularly because of the lack of sufficient experimental evidence. A subsequent paper on this subject[40] did not improve the understanding of either the polymer structure or the mechanism.

Ishigaki, Shono and Hachihama[36, 41] restudied the system using both Lewis and Brønsted acids. Their results indicated a fairly complex situation characterized by the formation of low molecular-weight products ($300 < \bar{M}_n < 1300$) the structure of which seemed to change with the polymerization temperature. However, their conclusions on the structure of these oligomers only made clear that previous simplistic assertions[37–39] were not reliable. These authors argued that their spectral and analytical evidence suggested the absence of furan rings in the polymers but the presence of both —CH=CH—O— and —CH=C(CH_3)—O— units. Hydroxyl groups were present in some polymers but absent in others. Finally, the formation of a tetramer of 2-methylfuran was observed when 2-methylfuran was refluxed with 3% phosphoric acid for 30 minutes. This latter reaction was examined in more detail in a recent publication by the same authors[42] and the structure of the tetramer established as:

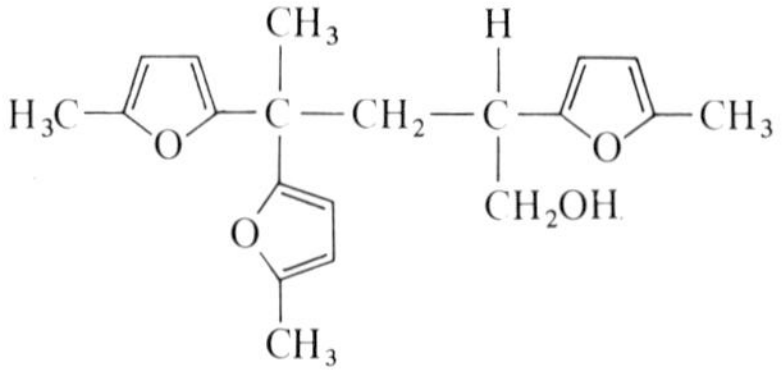

13

A mechanism for its formation was also proposed. Essentially, this involved protonation of 2-methylfuran followed by dimerization and trimerization to a 2,4-difuryl tetrahydrofuran derivative which suffered an acid catalysed cleavage of the saturated ring to produce a carbenium ion possessing an alcoholic function at the other end of a 5 carbon aliphatic chain. The latter finally added onto another molecule of

2-methylfuran, expelled a proton and generated the tetrameric carbinol *13*. The fact that not one of the intermediate neutral molecules postulated in this mechanism was isolated leaves the whole mechanism open to doubt. However, the isolation of the carbinol and the speculation that the oligomers obtained from that same system and others are in fact aliphatic-olefinic chain products with side 2-methylfuran rings, carbonyl and hydroxyl groups[42] is interesting and implies a striking difference between the behaviour of furan and 2-methylfuran in cationic polymerization.

Obviously much more mechanistic work needs to be done on this subject because the inconsistencies are too many and too serious. Thus, the latter authors proposed a mechanism for the oligomerization of 2-methylfuran by phosphoric acid[42] where water plays a dominant role in promoting ring cleavage ($+H_2O$) and in transforming a diol system into a ketonic one ($-H_2O$) yet, the formation of the tetramer discussed in the very same context is explained without the intervention of water. It is not clear from the experimental section whether the acid is used in a pure form or in aqueous solution. It is also unclear whether the tetramer postulated is present as such at the end of the reaction or if it is formed from a less stable entity during the products' isolation procedure, and it is surprising that the preparation of the tetramer was not attempted with other acidic catalysts because it leaves the reader with the impression that this reaction is peculiar to phosphoric acid.

2,5-dimethylfuran is more resistant to acids than furan and 2-methylfuran[36] and the reasons for its reluctance to polymerize are not obvious.

d) γ-Ray Irradiation. Irradiation of liquid furan with ^{60}Co γ-rays produces branched oligomers showing evidence of unsaturation, ether linkages and to a minor extent carbonyl groups[35, 43]. Similar results were obtained in the solid state and on furan adsorbed on carbon black[35, 44]. The radiolysis of the monomer precedes polymerization. Probably the fragments of this scission are responsible for initiation, but the subsequent oligomerization does not seem to be of a cationic nature[35]. However, the possibility of it being carried out by free radicals is equally unlikely because of the reluctance of furan to give chain reactions with such intermediates (see point a) above).

e) Stereospecific Systems. In 1959, Topchiev, Golfarb and Krenstel[45] claimed that furan and 2-methylfuran polymerized with Ziegler-Natta catalysts. Two years later[38] details were published about the conditions used: when the proportion of $TiCl_4$ in the catalyst combination with $AlEt_3$ was increased, the polymer yield increased and, with a molar ratio of 5:1 in favour of the Lewis acid, "partial resinification occurred". This was with furan and only one yield was actually reported for $TiCl_4$: $AlEt_3$ = 3 in hexane at 10 °C and 6 hours of reaction: 20% of solid polymer and about 10% of oligomers. With 2-methylfuran some data were actually given more explicitly and a plot of yield *vs.* catalysts ratio showed that virtually no polymer was formed when an excess of $AlEt_3$ was present. The yield increased from 10 to 20% when the excess of $TiCl_4$ was increased from 3:1 to 5:1. Also, it was shown that the yield increased as the temperature decreased. All this evidence clearly indicates that the polymerization is simply a cationic one initiated by titanium tetrachloride and also by the aluminium ethyl chlorides formed in the interaction between the two catalytic components, while the real "complex catalyst", which operates with an excess of the organometallic compound, is inactive. Thus the original claim is incorrect. Recent work has shown[46] that under a variety of conditions, typical of strong Ziegler-Natta catalysis, furan and 2-methylfuran are virtually inert and are recovered almost quantitatively after several hours' contact with the catalyst. Furan has in fact been used to enhance the stereospecificity of $TiCl_4-AlEt_3$ catalysts (1:2) in the polymerization of styrene[47]; the results were interpreted in terms of the furan inhibiting the cationic polymerization of styrene by the aluminium alkylchlorides through reacting or complexing with them. A similar study on ethylene-propylene copolymerization showed that furan enhances the time stability of the catalyst[48]. 2-furaldehyde has also been found to increase the stereospecificity of $TiCl_3-AlEt_2Cl$ catalyst in the polymerization of butene-1[49]. In none of these reports was the furan compound reported to undergo polymerization.

In conclusion, furan and 2-alkylfurans can be polymerized only by acidic initiators or by γ-radiation because the other standard methods of polyaddition fail to induce a chain-propagation reaction.

2. Copolymerization of Furan and Homologues with Maleic Anhydride

It is well known that a monomer which cannot homopolymerize often gives alternating copolymers with suitable co-monomers. Furan and its homologues display precisely this behaviour with electron-deficient monomers in radical polymerization. Butler, Badgett and Sharabash[50] reported in 1970 that furan and maleic anhydride form a donor-acceptor complex capable of giving an alternating copolymer when treated with azo-bis-isobutyronitrile in benzene solution. The composition of the copolymer was independent of the monomer feed ratio and close to 50/50; molecular weights were low (1000–2000). These authors pointed out that since both the Diels-Alder adduct and the complex between the two monomers were shown to give the same copolymer, the mechanism was not clearly resolved and the role of each species would have to be studied further. Gaylord and calloborators[51, 52] continued the investigation of this system and extended it to the 2-methylfuran-maleic anhydride pair. While confirming previous observations[50], they studied the spectra of the copolymers and proposed the following structures:

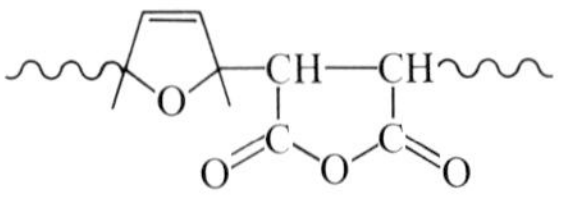

14   and   15

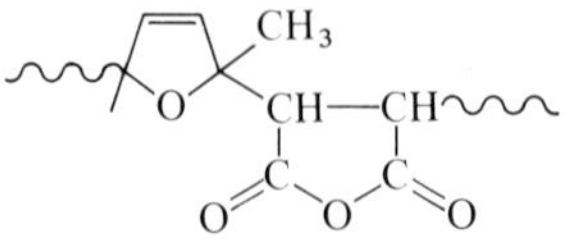

They also argued that the polymerization of the adduct proceeded only when the retrograde reaction giving back the two monomers was occurring, *i.e.* at higher temperatures and thus the real "monomer" was the charge-transfer complex.

This interpretation has been challenged by Kamo and co-workers[53] who independently studied the same systems and arrived at the conclusion that the polymer arises from the radical homopolymerization of the Diels-Alder adduct through its residual double bond, to give structure *16*.

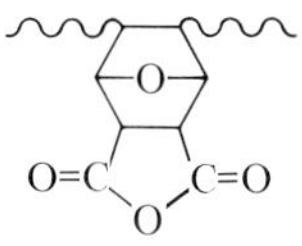

16

The close similarity of the polymers obtained by mixing the monomers and a radical initiator at 70 °C and by heating the preformed adduct at the same temperature with the same initiator was taken as major evidence in favour of the structure proposed. However, these authors[53] failed to consider that the retrograde reaction is important at 70 °C and therefore the two experiments are not discriminating. Also, in discussing the various alternative structures the polymer could have, they did not take into account the one in which the furan ring copolymerize through the 2,5-positions, i.e. the sequence *14* proposed by Gaylord et al.,[51].

The controversy has now been settled thanks to the studies of Russo and co-workers[54], and of Ragab and Butler[55]. The radiation-induced polymerization of furan and maleic anhydride mixtures at room temperature gave an oligomer the structure of which was again analysed spectroscopically and proved to be *14*[54]. The Diels-Alder adduct did not give any polymer under these conditions, which indicates that the retrograde reaction is responsible for the polymerization of the two monomers formed from this compound at higher temperatures. Ragab and Butler showed that the polymer contained C=C unsaturation in amounts close to the value expected from structure *14* (structure *16* does not possess any such unsaturation); the ^{13}C-NMR spectrum of the polymer was also fully consistent with structure *14*[55].

It now seems well established that the charge-transfer complex formed between furan and its homologues and maleic anhydride, probably in its excited state[51], is the species that polymerizes; the nature of the interactions in the complex being strong enough to overcome the restrictions to radical polymerization exhibited by furan derivatives. Presumably, other electron-deficient compounds could play the same role as maleic anhydride and other furan copolymers be prepared as indicated in the original work on this topic[50].

3. Benzofuran and Naphthofurans

Furan derivatives with an aromatic system fused on one of the ring's double bonds, such as benzofuran, naphthofuran etc., can be polymerized cationically through the other ring's double bond. In these polymerizations the complications encountered with furan and alkylfurans [see Section III-A-l-c] are absent because only one unsaturation is available for propagation, the other being "tied up" in the benzene system.

Thus, the polymers have a regular structure and can attain high molecular weights under controlled reaction conditions.

Benzofuran was first polymerized by Kraemer and Spilker[56] using different amounts of sulphuric acid or iodine. In that same period Stoermer[57] reported the tendency to polymerize of various methyl- and methoxy- substituted benzofurans. Many decades later the interest in this monomer was revived by Sigwalt[58] who showed that $TiCl_4$ at low temperature produced high-molecular-weight polymers with some chain branching and cross-linking probably due to alkylation of the benzene ring, and lower molecular-weight copolymers with styrene, α-methylstyrene, anethole, isobutyl-vinyl ether and indene. It was also noted that benzofuran polymerized less readily than indene in the same conditions. Okamura and co-workers[59] extended Sigwalt's work and found that the catalytic combination $SnCl_4 \cdot CCl_3COOH$ initiated the polymerization of benzofuran at 30 °C, while $BF_3 \cdot Et_2O$ failed to do so, providing another indication of the relative reluctance of this monomer to polymerize with cationic initiators. Copolymerization studies with styrene were also carried out and the values $r_1 = 0.80$ (styrene) and $r_2 = 0.95$ (benzofuran) obtained at 30 °C; this seems to indicate a higher incorporation of benzofuran in the copolymer than expected from the much lower homopolymerization rate of this monomer compared with that of styrene. This apparent contradiction was explained in terms of an artificially high concentration of solvating benzofuran molecules around the growing ion pairs which would favour the addition of this monomer to a higher degree than its relative reactivity would normally allow.

An interesting aspect of the benzofuran cationic polymerization was uncovered by Natta, Farina, Peraldo and Bressan who reported in 1961[60, 61] that an asymmetric synthesis of an optically active poly(benzofuran) could be achieved by using $AlCl_2Et$ coupled with (−)β-phenylalanine, (+)camphorsulphonic acid or with (−)brucine. The optical activity was definitely due to the asymmetric carbon atoms in the polymer chain, indicating that at least some of the polymer's macromolecules possessed a di-isotactic structure, viz.[62]:

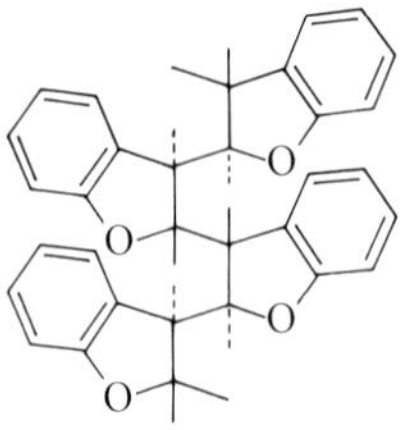

17

These polymers could not be crystallized, despite their apparent stereoregularity, probably because of the sterically-hindered character of the chains. It was proposed by Farina and Bressan[62] that the chain growth was stereoregulated by the optically active anion of the ion-paired chain carrier. Further studies[63] showed that the first portion of the polymer produced in a given reaction always possessed a less regular structure than later portions, unless the reaction was started in the presence of previously prepared polymer. This observation was interpreted as evidence for the presence of two types of active species at the beginning of the polymerization, one essentially non-stereospecific derived from uncomplexed Lewis acid, and the other stereospecific, derived from the complexation of the Lewis acid with the optically active co-catalyst; the free acid in excess would progressively lose its activity by complexing with the polymer formed and the system would correspondingly acquire a more discriminating character in terms of orienting power upon the incoming monomer. Panton and Plesch[64] confirmed this interpretation when they showed that the presence of water in these systems played a negative role because it competed with the optically-active co-catalyst in forming chain carriers, thus reducing the stereospecificity of the system.

Experiments with a different catalyst system, $SnCl_4$–$AlEt_3$-(–)menthol, carried out in another laboratory[65] gave strength to the hypothesis that the counteranion is a dominating factor in the stereoregulated growth if the chain carrier is an ion pair.

Recently, Bressan and Broggi[66] succeeded in crystallizing poly(benzofurans) obtained with cationic catalysts. Surprisingly, both optically active and inactive polymers could be crystallized, although an important difference was found between them: the latter could not be crystallized from solutions. It now seems probable[67] that all polymers prepared with aluminium-based Lewis acids possess a certain degree of stereoregularity even if the counterion is not optically active and that the lack of optical activity in the latter case arises from interchain racemization or from insufficient stereospecificity (sufficient, however, to allow crystallization under certain conditions). The crystalline polymers melted at 355–360 °C with some decomposition[66].

Further work on these systems would obviously be very welcome, considering the implications of such stereoregulated mechanisms to biological systems, and in view of the rather unexpected findings reported lately[66, 67] which add new interest to this field.

The behaviour of naphthofurans was found to be entirely similar to that of benzofuran[68].

4. Dihydrofurans

Strictly speaking, dihydrofuran compounds do not belong to the field covered by this review. However, their behaviour in polymerization reactions is both similar and "cleaner" than that of their furanic counterparts and it is felt that their brief inclusion here makes the panorama more complete and perhaps clearer in some respects.

Neither 2,5-dihydrofuran nor 2,3-dihydrofuran polymerize with free-radical initiators and it has been reported that 2,5-dihydro-2,5-dimethoxyfuran retards the radical polymerization of vinyl acetate[69]. 2,3-dihydrofuran has been found to undergo radical and spontaneous copolymerization with maleic anhydride and acrylonitrile to give the classical alternating copolymers arising from the establishment of a 1:1 charge-transfer complex between the monomers[70]. Radical copolymers have also been claimed[70] with styrene and methyl acrylate but in these systems the reactivity ratios favour the olefinic co-monomers by about 20 to 1 making the copolymers poor in cyclic ether content and the rates of polymerizations slower the higher the 2,3-dihydrofuran proportion in the monomer feed ratio. It seems likely that the latter situations, where no complex is formed between the monomers, reflect a retardation phenomenon with corresponding decrease in molecular weight; therefore most of the oxygen content in the polymers, ascribed to tetrahydrofuran units, is probably in the end groups, both from the termination by the cyclic ether and from the dibenzoyl peroxide fragments.

It is known from the classical work of Barr and Rose[71] that, while 2,5-dihydrofurans do not polymerize with cationic initiators, 2,3-dihydrofurans do and give linear polymers with tetrahydrofuran rings as repeat units. Schmitt and Schuerch[72] studied the cationic polymerization of 2,3-dihydro-5-methylfuran in the presence of optically-active Lewis bases as co-catalysts to test the possibility of stereospecific growth. In fact, the polymers did not show any optical activity and the authors proposed the growth of a chain through *trans-trans* additions to give an internally compensated structure. Kamo and co-workers[73], however, succeeded in synthesizing stereoregular polymers with this monomer by using Ziegler-Natta catalysts or standard Lewis acids at low temperature. Optical activity was observed in polymers prepared with Lewis acid optically-active Lewis base combination, indicating that erithro-di-isotactic chains had been formed in the polymerization. This study is reminiscent of similar work on the polymerization of benzofuran (see Section III-A-3); the asymmetry of both carbon atoms of the repeat unit in the polymer chain gives rise to a potential for stereospecific growth to give optically active products. It is interesting to note that stereoregular poly(2,3-dihydro-5-methylfuran) does not crystallize readily[73], just like its poly(benzofuran) counterpart.

A comparison of the cationic polymerization of 2,3-dihydrofurans with that of furan and 2-alkylfurans shows that the complications of the latters two, arising from the dienic character of the monomers, obviously vanish when the monomer is a simple cyclic vinyl ether with just one reactive site, *viz.* the carbon-carbon double bond. However, it also points out that ring opening in the polymerization of furans by acidic catalysts in the absence of water is unlikely, because otherwise it would also occur to some degree in the polymerization of dihydrofurans.

The only known instance of ring-opening polymerization with these compounds is also the only report on the successful polymerization of 2,5-dihydrofuran[74] in which this compound was cationically copolymerized with epichlorhydrin ($r_1 \sim 0$, $r_2 \sim 0$), propylene oxide ($r_1 \sim 0, r_2 \sim 0$) and 3,3-bischloromethyl oxacyclobutane ($r_1 \sim 0, r_2 = 1.6$). It was shown that all the copolymers obtained possessed a certain degree of unsaturation which was attributed to the presence of open units from 2,5-dihydrofuran. Thus, for example, the alternating copolymer with epichlorhydrin had the following structure (IR spectra, Cl content, C=C analysis):

$$-(O-CH_2-CH)-(O-CH_2-CH=CH-CH_2)-$$
with CH$_2$Cl substituent on the first CH.

5. The Photopolymerization of 2-Furaldehyde

Among the various agents which provoke the resinification of 2-furaldehyde, ultraviolet light has been the least studied. Some comments have already been made on the action of heat and acids on this molecule in an inert atmosphere; the role of oxidizing agents on the "destruction" of 2-furaldehyde is well documented, although the mechanism of these reactions is not fully understood because of its extreme complexity.

While the products of 2-furaldehyde polymerization by heat are branched polycondensates with highly conjugated structures (see Section II-C), the photopolymerization of this furan derivative gives a linear polyaddition product [24,75].

The gas-phase photolysis of 2-furaldehyde in the $\pi^* \leftarrow n$ and $\pi^* \leftarrow \pi$ transitions [76] proceeds with fragmentation to CO, furan and C_3-hydrocarbons, but a certain amount of resinification is also noted (about 5% quantum yield with excitation of the $\pi^* \leftarrow n$ transition). The latter observation prompted a study of the vacuum liquid-phase photolysis by sunlight or by light from a medium-pressure mercury arc at room temperature [24, 75]. The resin obtained was submitted to fractionation and structural analysis. On the basis of the results obtained and other mechanistic evidence, the following sequence of events was postulated for the photopolymerization:

— the electronically excited 2-furaldehyde molecule reacts with a ground-state molecule to give vibrationally excited dimers:

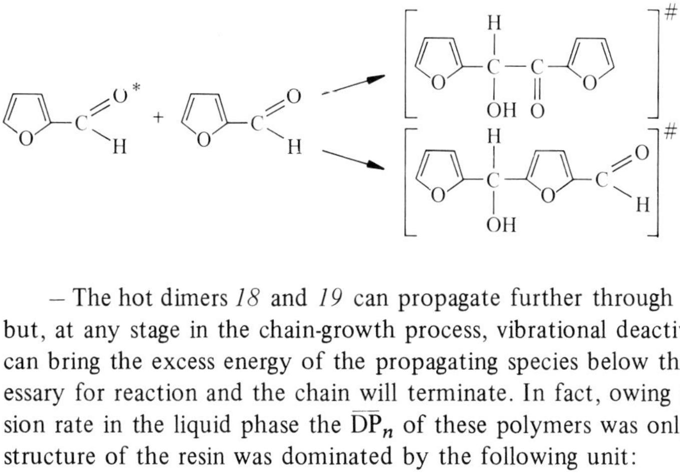

— The hot dimers *18* and *19* can propagate further through the above reactions; but, at any stage in the chain-growth process, vibrational deactivation by collision can bring the excess energy of the propagating species below the critical value necessary for reaction and the chain will terminate. In fact, owing to the high collision rate in the liquid phase the $\overline{DP}_n$ of these polymers was only about 5. The structure of the resin was dominated by the following unit:

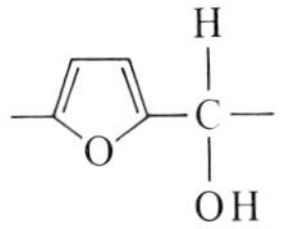

20

The presence of one carbonyl group *per* oligomer molecule was also ascertained. The orange colour of the resin suggested that some minor event during the photopolymerization produced chromophores in small concentrations. The presence of furoin among the products corroborated the proposed mechanism, which was shown not to involve free radical chain reactions.

6. Miscellaneous Reactions

In this section some systems are briefly described which make use of furan derivatives as starting materials for chemical transformations which eventually lead to polymeric products. However, with the exception of Section (c), these final products bear little resemblance to the original furan compounds used and are, therefore, only marginally relevant to the context of the present review.

a) The Paterno-Büchi Reaction with Furans. Andrews and Feast[77] have studied the possibility of preparing polymers by photolysing furan and 2,5-dimethylfuran with aromatic diketones in benzene solution. They hoped that photo-cycloaddition of the carbonyl group onto the furan double bonds (a well-established reaction) would give a polymer with the following structure:

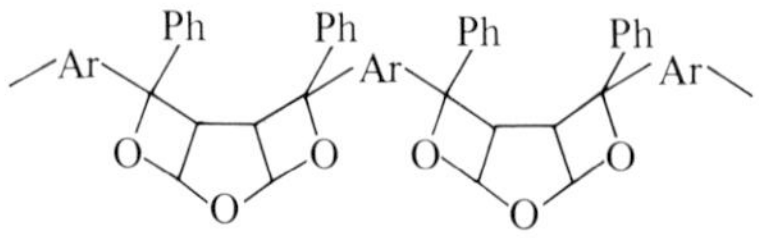

However, molecular weights were low because of the difficulty of maintaining a precise stoichiometry during the reaction. The preparation of 2:1 solid furan-diketone adducts and their subsequent irradiation in solution in the presence of equimolar quantities of an aromatic diketone made it possible to overcome this problem and polymerization occurred. It was noticed that the rates of these polymerizations were low, several days of irradiation being necessary for the attainment of a maximum, constant molecular weight ($\bar{M}_n \sim 5000-10{,}000$). Most of the systems were marred by side reactions leading to branching and some cross-linking. Despite these drawbacks one must consider that this is the first attempt to use furan derivatives as starting materials in a new type of polymerization process and hope that future work might give a better understanding of these systems and therefore, better ways of achieving the initial goal.

b) The Aromatization of bis-Furan Adducts. Berlin and co-workers[78] have recently reported the preparation of temperature-resistant aromatic polymers derived from Diels-Alder adducts involving *bis*-furans. The reaction pathway consists in preparing the *bis*-furan derivative, making its double adduct with maleic anhydride, dehydrating the latter to give an aromatic dianhydride and finally heating this with aromatic di- and tetraamines to obtain poly(aroylen-*bis*-benzoimidazoles) or polyimides. It is quite apparent that the role of the furan derivatives is purely synthetic, since literally no trace of furan character is to be found in the final products.

c) Poly(2,5-difurylketone). An interesting reaction between carbon monoxide and 2,5-dichloromercurifuran was described by Izumi, Iino and Kasahara[79]. Under a CO pressure of 50 atm lithium chloropalladite catalyses the formation of a polymer of the type:

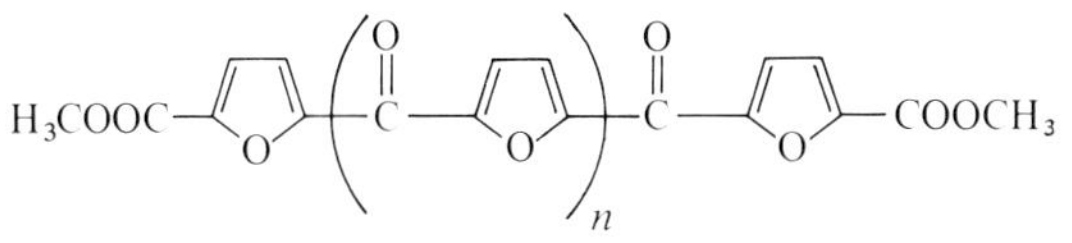

[Structure: H₃COOC–furan–(C(=O)–furan)ₙ–C(=O)–furan–COOCH₃]

(terminal groups due to methanol, used as solvent) with a $\overline{DP}_n$ of about 12.

B. Polymerization through a Function External to the Ring

1. 2-Alkenylfurans

The monomers considered in this section possess an olefinic bond conjugated to the ring through the C-2 position. The parent compound, 2-vinylfuran, and some of its methylated homologues are probably the best understood monomers of the furan series in terms of their behaviour and peculiarities in different polymerization systems. This consideration is of course relative, *i.e.* the knowledge acquired on these compounds in recent years allows the formulation of certain conclusions with a high degree of confidence and this is already exceptional in the field covered by this review.

Considerable interest was placed on 2-vinylfuran in the years 1930–1950 and a number of patents and publications on its polymerization and copolymerization appeared. This work was aimed at exploring the potential of the new polymer and concentrated therefore, on descriptive aspects of its preparation, properties and possible applications (for bibliographies on this topic, see[3, 80, 81]). Little was done to shed light on the more fundamental aspects of the systems studied until about 20 years ago. The present publication reviews these later studies.

a) Free-Radical Systems. 2-Vinylfuran polymerizes thermally in the absence of oxygen (110–150 °C) with an overall activation energy of 17 kcal/mole and an external order in monomer close to three[82]. While these features resemble those obtained with styrene, some peculiarities are encountered, namely, the molecular weights of the polymers are rather low ($\overline{M}_n$ = 3000–8000), the yield-*vs.*-time curves tend to level off at fairly low conversions and an important side reaction is observed which gives the rearranged, Diels-Alder dimer[82, 83]:

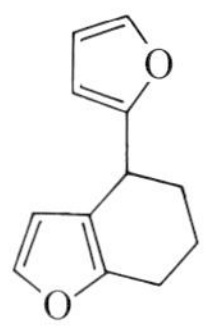

Similar anomalies have been encountered by several workers in the bulk and solution polymerization of this monomer induced by classical free-radical initiators[84–86]; also, particularly low rates of conversion were observed. The most thorough kinetic study was carried out by Aso and Tanaka[86] who again found "normal" results and a value of k_p^2/k_t much lower than that for styrene. Copolymerization studies of 2-vinylfuran (M_1) have given the following values of the reactivity ratios:

2-vinylfuran/styrene	$r_1 = 1.9$, $r_2 = 0.25$ [86];
2-vinylfuran/acrylonitrile	$r_1 = 0.82$, $r_2 = 0.037$ [86];
2-vinylfuran/vinylidene chloride	$r_1 = 11.7$, $r_2 = 0.15$ [87];
2-vinylfuran/butadiene	$r_1 = 4.52$, $r_2 = 0.11$ [88].

All this evidence suggests that the radical produced from 2-vinylfuran is a rather strongly stabilized entity, compared with those of more common monomers, and is therefore, not very active in homopolymerization. On the other hand, because of its relative stability, it does not add easily to monomers like styrene, vinylidene chloride or butadiene, and thus the copolymerization rates are also low. Aso and Tanaka[86] calculated the values of Q and e as 2.0 and 0.0, respectively.

Emulsion polymerization of 2-vinylfuran gives high yields and molecular weights[85, 31] in contrast with the bulk and solution systems. This technique is undoubtedly the best yet found for the synthesis of poly(2-vinylfuran).

The photopolymerization of this monomer with a mercury arc[89, 90] produces small yields of low molecular-weight products. In the presence of oxygen an induction period is noted and the polymers contain an appreciable amount of peroxide units in the chains[91]. The photolysis of 2-vinylfuran was briefly described by Hiraoka[92]: cyclopentadiene and CO were reported as products. It is not certain if free radicals are involved in this photodecomposition, but presumably they are.

While a careful spectroscopic investigation of the structure of poly(vinylfuran) obtained by radical initiation[93, 94] leaves no doubt about the predominance of the normal vinyl-type units in its chains, some discussion is needed in order to attempt an interpretation of the abnormally low rates, yields and DP's obtained. It is known that certain furan derivatives play a retarding or inhibiting role in radical polymerization (see Chapter V) and it is conceivable that small concentrations of by-products, particularly at the higher temperatures or with ultraviolet light, would accumulate during the radical polymerization of 2-vinylfuran and subsequently act as chain-termination agents. Alternatively, it could be argued that the poly(2-vinylfuran) formed acts itself as a retarder through the combined action of its many furan rings, in the same way as 2-methylfuran retards certain radical polymerizations (see Chapter V). Obviously some mechanistic investigation of this problem is needed, particularly in view of the difference of behaviour between emulsion and homogeneous media.

5-Methyl-2-vinylfuran behaves like its parent monomer in radical polymerization[95] but the yields and molecular weights of the polymers are somewhat higher.

2-Isopropenylfuran resists radical homopolymerization[96, 97] except when high pressures are used[97], but even then low molecular-weight products are formed. Radical copolymerization with several monomers is, however, possible[96, 98]. 5-Isopropyl- and 5-methyl-2-isopropenylfuran behave similarly[95, 96]. These results are not unexpected for 1,1-disubstituted vinyl monomers and are almost certainly connected with thermodynamic restrictions, *i.e.* fairly low ceiling temperatures. If this is true, photosensitized polymerizations at low temperatures should give homopolymers with these monomers.

b) Anionic Systems. Conflicting reports about the interaction of 2-vinylfuran with anionic initiators have appeared in the literature. Andreeva and Koton[99] reported that this monomer polymerized with butyl lithium but gave no details. Goutière, Léonetti and Golé[100] claimed that sodium reacts with 2-vinylfuran in tetrahydrofuran to give the corresponding radical anion which then dimerizes to the dianion and this propagates at room temperature to give poly(2-vinylfuran). The only detail given was the formation of a red-orange colour when monomer and catalyst were mixed. It was also claimed that a block copolymer with methyl methacrylate had been prepared; again, no evidence was presented. In another publication[31] these authors reported the polymerization of 2-vinylfuran with sodium-naphthalene in the same solvent at room temperature and a similar mechanism was proposed based on the colour change produced when the catalyst was added to the monomer. However, in this case it was also stated that the polymer formed had a molecular weight compatible with a classical living-polymer situation. The above results are in marked contrast with previous observations published by Kuwata, Kawazura and Hirota[101] that the reaction of 2-vinylfuran with sodium in tetrahydrofuran (all-glass apparatus, high-vacuum) gives a species with a maximum at 540 nm which is active in the initiation of acrylonitrile and methyl methacrylate polymerization, but does not give polymer with 2-vinylfuran. This intermediate was found to give an EPR signal the intensity of which decreased with time. Both the visible absorption and the EPR signals vanished when air was admitted into the reactor and the ultraviolet spectrum of the resulting solution indicated the presence of conjugated products. The authors proposed an oligomeric radical anion but did not speculate on its fate with time or in the presence of (moist) air.

Gandini and Hernandez[46, 102] recently found that:

— 2-Vinylfuran did not polymerize with butyl lithium in hexane or tetrahydrofuran. Traces of resinous materials were isolated. Their spectra indicated complicated structures and not poly(2-vinylfuran).

— When undiluted 2-vinylfuran was added to metallic sodium (mirror or particles) an orange colour developed and some resinous material was deposited on the metal surface. On prolonged contact much of the monomer was converted into a partly-insoluble reddish resin with spectra unrelated to those of standard poly(2-vinylfuran). Reaction of diluted monomer with sodium gave a milder interaction, but no evidence of living anionic polymerization.

— The reaction of 2-vinylfuran with sodium naphthalene in tetrahydrofuran under controlled conditions (high vacuum manipulation of very pure and dry reagents) over a wide range of catalyst concentrations produced the following phenomenology: the green colour of the naphthalene anion disappeared instantly upon adding the monomer; only when the catalyst concentration was high ($10^{-2}-10^{-1}$ M) was a slow colour build-up noticed and recorded with a spectrophotometer. Within a day the solution was neutralized, a spectrum taken and products sought. 2-Vinylfuran was recovered in almost quantitative amounts, except for minute traces of a residue which could not be analysed. The spectra of the solutions before and after neutralization were similar to those reported by Kuwata, Kawazura and Hirota[101].

The disagreement between these findings and those of Andreeva and Koton and Goutière, Léonetti and Golé is striking. The author of the present review is inclined to think that the material isolated by Andreeva and Koton was a resin similar to those obtained by Gandini and Hernandez with some of their systems, and that their structure must involve the participation of the furan ring as well as the vinyl double bond. The interaction of furan and 2-ethylfuran with anionic initiators [see Section III-A-1-b)] supports this interpretation. As for the work of Goutière, Léonetti and Golé, it is odd that these authors described in the very same papers[31, 100] the oligomerization of 2-ethylfuran and the cross-linking of poly(2-vinylfuran) by alkali metals and sodium naphthalene, giving details of colour of both the reacting solutions and the oligomers, but offered no comment on the possible participation of the furan ring when 2-vinylfuran is "polymerized" under the same conditions. The lack of any experimental evidence as to the structure of both homopolymer and block-copolymer claimed; the revealing statement that the "radical anion" (or the dianion) of 2-vinylfuran is sparingly soluble in tetrahydrofuran; the fact that the polymerization was said to be slow and yet the concentration of sodium naphthalene was abnormally high; the occurrence of apparent inconsistencies such as the statement that the monomer was twice distilled over sodium[31], while in a preceding report sodium and the monomer were reacted to immediately give the carbanion which induced the polymerization at room temperature[100], are all strong elements which, when combined with present evidence[102], cast serious doubts on the claim that 2-vinylfuran can undergo anionic living polymerization.

It can be concluded that this monomer does not polymerize with standard anionic initiators, but that minor side reactions involving the vinyl group and the furan ring give rise to small yields of resinous materials of complex structure. It was recently noticed[93] that both 2-vinylfuran and 5-methyl-2-vinylfuran kept over finely powdered CaH_2 in the absence of air or moisture, slowly produced similar resinous cross-linked materials which are also orange. The rapid reaction between 2-vinylfuran and sodium naphthalene is certainly worth studying, because the fate of the species formed when the electron is transferred to the furan compound could explain its failure to induce polymerization through the vinyl bond.

c) Cationic Systems. 2-Alkenylfurans readily polymerize with cationic initiators such as Lewis and Brønsted acids and iodine. However, before Alvarez, Gandini and Martinez started a systematic investigation of the kinetics and mechanism of these processes, only one publication had dealt quantitatively with this topic. All other previous work was purely documentary, although Aso and Tanaka[86] had recognized that the polymers obtained with $AlEt_2Cl$ possessed a complicated structure.

Stoicescu and Dimonie[103] studied the polymerization of 2-vinylfuran with iodine in methylene chloride between 20 and 50 °C. The time-conversion curves were not analysed for internal orders but external orders with respect to catalyst and monomer were both unity. Together with an overall activation energy of 2.5 kcal/mole for the polymerization process, these were the only data obtained. Observations about the low DP's of the products, their dark colour, their lack of bound iodine and the presence of furan rings in the oligomers, inferred by infrared spectra (not reported), completed the experimental evidence. The authors proposed a linear, vinylic structure for the polymer, and a true cationic mechanism for its formation and discussed the occurrence of an initial charge-transfer complex on the basis of the changes of colour of the polymerizing solution. Gandini, Martinez and Sanchez restudied this system briefly[104] and found it to be much more complex than indicated by Stoicescu and Dimonie. The oligomers produced were in fact both alkylated at C-5 and polyunsaturated (colour), just like their counterparts prepared with trifluoroacetic acid and other cationic initiators. The origin of these anomalous structures is discussed below.

The recent work of Alvarez, Gandini and Martinez[90, 93, 94, 105–107] is now summarized. The course of 2-vinylfuran polymerization catalysed by various cationic initiators (particularly trifluoroacetic acid) in methylene chloride at $-78\,°C$ to $40\,°C$ was carefully followed using various techniques. The structure of the products was analysed in detail and to elucidate some mechanistic problems various model reactions were investigated. This approach, extended to three other 2-alkenylfurans appropriately chosen for their strategic methyl substitutions, was the key to the understanding of a complex phenomenology where the furan ring plays an active and peculiar role superposed on the classical reactions of the olefinic double bond. Essentially, three major reactions dominate the cationic polymerization of 2-vinylfuran after initiation has taken place:

— propagation through the vinyl double bond;
— alkylation at the C-5 position of the ring of both monomer and polymer by an active species; this alkylation can also take place in an anti-Markovnikov fashion;
— hydride transfer from an unsaturated polymer molecule to an active species to give allylic-type carbenium ions in equilibrium with polyconjugated molecules.

Thus, poly(2-vinylfurans) prepared by these systems possess a variety of units as shown below:

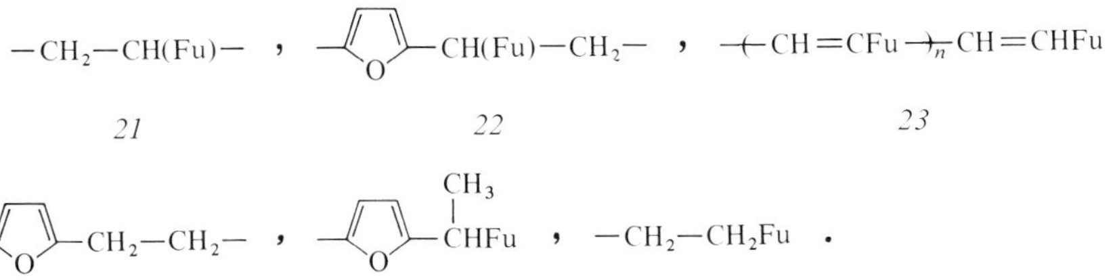

$-CH_2-CH(Fu)-$, ⌬$-CH(Fu)-CH_2-$, $+CH=CFu\xrightarrow{}_n CH=CHFu$

 21   22   23

⌬$-CH_2-CH_2-$, ⌬$-\overset{CH_3}{\underset{|}{C}}HFu$, $-CH_2-CH_2Fu$.

 24   25   26

These structures occur in branched chains of low DP. Particularly interesting is unit 24 which predominates in polymers prepared at $-20\,°C$ with trifluoroacetic acid; these products are in fact phantom polymers[108] in which everything seems to have gone the wrong way.

Elementary considerations indicate that with appropriate substitutions some of the reactions mentioned above can be eliminated. Indeed, when 5-methyl-2-vinylfuran was used, no alkylation was observed, the positions C-3 and C-4 being rather unreactive[16], and the polymer was a mixture of linear chains with polyunsaturations and linear saturated chains, i.e. only structures like 21, 23 and 26 were present, with a 5-methyl ring instead of the 5-unsubstituted one. When 2-isopropenylfuran was used, no hydride transfer took place since this requires a hydrogen atom in the α-position to the ring, which this monomer does not have; the polymers were white and gave electronic spectra transparent down to 280 nm. Alkylation at C-5, however, was an important feature at the higher temperature, and these poly(2-isopropenylfurans) showed units like *21, 22* and *25* with a methyl group instead of the α-hydrogen. At about −78 °C the propagation through the vinyl bond prevailed and the polymers had reasonably high molecular weights, with practically the normal structure *21-αMe*. 2-Isopropenyl-5-methylfuran was also studied; around room temperature oligomerization with formation of a fair amount of unsaturated dimer was observed, due to the low ceiling temperature of this monomer. Note that with 2-isopropenylfuran higher oligomers are obtained at these temperatures because the concurrent incidence of propagation through the double bond and of alkylation is effectively giving a copolymerization situation and thus the thermodynamic restrictions do not apply. At lower temperatures the polymers were linear vinyl chains which bore no indication of "side" reactions:

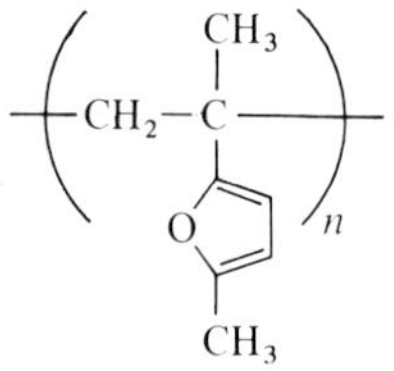

27

In order to ascertain the validity of the assignments made, particularly for the more complex structures, model compounds were synthesized and used as references.

The kinetics of these polymerizations were also thoroughly studied. It was established that the alkylation reaction could be treated as an alternative propagation step. Thus, the systems involving 2-vinylfuran and 2-isopropenylfuran were best represented as copolymerizations where instead of two different monomers one had to consider the same monomer with two alternative reactive sites. The hydride transfer reaction could be studied independently by visible spectroscopy and electrical conductivity. The formation of carbenium ions of considerable stability during the polymerization led to a progressive decrease in the concentration of acid available for initiation and was therefore treated as a termination step. This was confirmed by the high internal orders found in the polymerizations of 2-vinylfuran and 2-vinyl-5-methylfuran, and by following the rate of polymerization after a second addition of monomer. Another minor termination reaction common to all systems was also found from kinetic measurements and its origin was attributed to the formation of inactive ester molecules with the ring double bonds.

This investigation was complemented by a series of experiments where the carbinol precursors of the four monomers were used as "monomers" in the presence of trifluoroacetic acid or its anhydride. The results were similar to those obtained in an earlier study with aromatic carbinols[109] and indicated that one of the active species present in the polymerizations of the 2-alkenylfurans must be the ester derived by the initial reaction of the vinyl bond with the acid:

$$\underset{Fu}{\overset{R}{\diagdown}}C=CH_2 + CF_3COOH \rightleftharpoons CH_3-\underset{Fu}{\overset{R}{\underset{|}{C}}}-O-\overset{O}{\underset{CF_3}{\overset{\|}{C}}}$$

The effect of common-anion salts and of added water showed, however, that ionic chain carriers must also be present in these systems. These observations, coupled with experiments where dilute solutions of the monomers were treated with excess of acid and the reactions followed by ultraviolet spectroscopy, produced sufficient information about the initiation reaction pattern and thus completed the overall kinetic and mechanistic approach.

d) γ-Ray Irradiation. Only 2-vinylfuran has been studied by this technique[110]. A substantial proportion of cross-linked material was obtained with 2×10^7 r, clearly indicating that ring participation, probably following scission[43], is important. The infrared spectra of these polymers were not very informative, but definitely showed carbonyl and C=C bands, obviously coming from cleaved rings.

e) Stereospecific Systems. Before the recent work of Gandini and Hernandez[46, 102] there was only a single brief mention of the use of Ziegler-Natta catalysis for the polymerization of 2-isopropenylfuran[97] in the literature. An insoluble, infusible white polymer was obtained with $AlEt_3-TiCl_4$ (3:1) in heptane at low temperature.

Gandini and Hernandez found that 2-vinylfuran and some of its methylated homologues readily polymerized with a wide variety of catalyst combinations and classified the results according to the following three categories:

– With "strong" catalytic combinations such as $AlEt_3-TiCl_4$, $AlEt_3-TiCl_3$, etc. polymerization in heptane and toluene at 0–50 °C produced cross-linked polymers. Obviously, the furan ring participated in the polymerization. However, this participation was modest, since the polymers, although insoluble, swelled very considerably in solvents like xylenes and tetrahydronaphthalene, indicating that the networks were fairly loose. It was also noticed that the first polymer increments were soluble and only at higher conversions did the products lose solubility and acquire a brownish colour. It is conceivable that cross-linking could be promoted by occluded catalyst in the precipitated polymer, *i.e.* it cannot be ruled out that during the actual chain growth there was little or no side reaction involving the ring, but that the catalysts operated on the polymer formed (still suspended in the reaction medium or after its isolation and exposure to the atmosphere) attacking some of its rings and ultimately producing a gel.

– With "medium-strength" catalysts such as $BuOAlEt_2$ and $Al(isoBu)_3$ with various vanadium and titanium halides and oxyhalides, the polymers formed were white powders of moderate molecular weight, soluble in chloroform. High yields were obtained within a few hours at room temperature. These polymers are presently being studied to see if they possess any degree of stereoregularity and if so whether they can be brought to crystallize.

— With weak catalysts such as aluminium alkoxides with $COCl_3$, cobalt acetyl acetonate, etc. polymerization was exceedingly slow.

This work satisfactorily explored the field in terms of its potential, covering a fairly wide area. It now needs a more quantitative approach with those systems which gave good results and might prove stereospecific. Although the polymers were analysed in some detail, many properties need to be studied; for example, the relationship between intrinsic viscosity and molecular weight has never been established for any poly(alkenylfuran), nor has the glass transition temperature been measured. This type of work is now in progress.

2. Other Olefins Containing the Furan Ring

Sporadic mentions of vinylfuran derivatives and their polymerization are encountered in the literature. Thus, 5-nitro-2-vinyl furan was polymerized by free-radical initiators to give low molecular-weight products[111]. Koton and co-workers[84, 99, 112] studied the polymerization of 2-vinylbenzofuran, 2-vinyldibenzofuran and 3-vinyl-2,5-diphenylfuran and established some criteria for their relative tendency to polymerization with radical and cationic initiators. Bachman and Heisey[96] polymerized 2-isopropenylbenzofuran with benzoyl peroxide and discussed the possible structures of the polymers considering the monomer as a conjugated diene. Topchiev and Alaniya[113] reported the synthesis and polymerization of β-nitro-2-vinylfuran.

3. Vinyl Esters of the Furan Series

Hopff and Lüssi[114] and Hardy and Szita[115] were the first to synthesize vinyl 2-furoate and to notice its reluctance to undergo free-radical polymerization. Later, Hardy[116] showed that the radical polymerization of vinyl acetate was strongly retarded by both vinyl 2-furoate and phenyl 2-furoate and argued that the furan ring was responsible for these anomalies. Spektor[117] prepared a number of vinyl esters of the furan series and tested their polymerizability with radical and cationic initiators. Apart from some resinification with $BF_3 \cdot Et_2O$ and $FeCl_3$, these tests gave negative results. Copolymerization with methyl methacrylate was reported, but the products were not characterized. Spektor did not discuss his results. Recently, Kamo and co-workers[118] prepared oligomers of vinyl 2-furoate by treating it with up to 25% of azo-bis-isobutyronitrile. The structural analysis of these products, based on infrared and NMR spectra, was taken as evidence of a normal growth through the vinyl bond.

Gandini and Rieumont[26, 119] have carried out an extensive examination of the polymerizability of several vinyl esters of furan carboxylic acids and of the causes of the autoinhibition which most of them display with free-radical initiation. The compounds studied were the vinyl esters of 2-furoic, 2-furylacetic, 2-furylpropionic, 2-furylacrylic and sorbic acid. All these derivatives, showed the same strong indifference towards radical polymerization. Only when treated with large doses (10–30%) of initiator did they give small yields of oligomers. The structure of all these products was carefully studied by spectroscopic and other techniques. Invariably, it was noticed that an appreciable concentration of vinyl groups was still present in the products which did not decrease after several reprecipitations. Moreover, the NMR spectra showed evidence of dihydrofuranic structures except, of course, in the case of vinyl sorbate. More specifically, with vinyl 2-furoate the oligomers gave an infrared spectrum with a band at 1640 cm^{-1} characteristic of the monomer's vinyl stretching and the NMR spectrum showed signals between 4 and 6 ppm typical of the vinylic =CH_2 and of dihydrofuranic protons. These spectra were analysed in conjunction with signals from the monomer, isopropyl 2-furoate and a synthetic poly(vinyl 2-furoate) prepared from poly(vinylalcohol) and 2-furoyl chloride. The spectra of the oligomers obtained by Kamo and co-workers[118] coincided with those obtained by Gandini and Rieumont and thus the structural interpretation given by the former authors is incorrect, since they did not take into account the presence of vinyl and dihydrofuran groups. The tendency towards autoinhibition of all the esters used indicates that the phenomenon cannot be ascribed specifically to the particular structure of vinyl 2-furoate, i.e. the position of the vinyl-ester group with respect to the ring is not a critical factor. Moreover, the similarity of behaviour between the furan esters and vinyl sorbate seems to suggest that in this context the furan ring plays the role of a dienic moiety.

Several radical copolymerizations of vinyl 2-furoate with well-known monomers (50:50) were also studied. Complete inhibition was obtained with vinyl acetate, very strong retardation with styrene, vinyl chloride and acrylonitrile; methyl methacrylate homopolymerized without appreciable decrease in rate. It is evident that the degree of retardation that vinyl 2-furoate imposes upon the other monomer depends on the stability of the latter's free radical. With styrene and vinyl chloride the small amounts of fairly low molecular-weight products contained units from vinyl 2-furoate which had entered the chain both through the vinyl bond and through the ring (infrared band at 1640 cm^{-1}).

Vinyl 5-methyl-2-furoate slowly polymerized to give a polymer of *normal* structure when it was heated to 70 °C with radicals initiators. This observation is of considerable importance to the formulation of a general mechanism for the autoinhibition effect. It seems plausible to propose that the free radicals present in these systems are scavenged by the furan "monomer" through a reaction at the C-5 position. This leads to the formation of stable furylic radicals which have a very limited capacity to propagate; when they do propagate, they can either add to a vinyl group or to another furan ring thus giving rise to oligomers containing furan and dihydrofuran rings and vinyl bonds:

28   *29*

The presence of a methyl group at C-5 decreases somewhat the scavenging power of the ring and makes conditions more favourable for normal propagation. Work is presently in progress to confirm this point, using vinyl 2-furoates with bulky substituents at C-5.

An interesting development in this work was obtained when Gandini and Rieumont tried to activate the polymerization of vinyl 2-furoate with $ZnCl_2$. It was found that oligomerization occurred and the products had a structure which contained a substantial amount of alkylated units of the type:

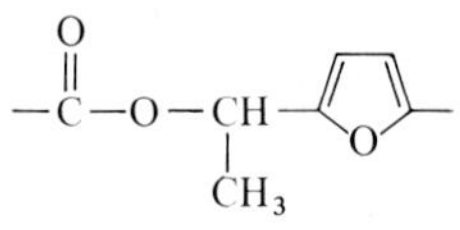

$$-\overset{O}{\underset{\|}{C}}-O-\underset{\underset{CH_3}{|}}{CH}-\text{[furan ring]}-$$

30

Thus, a polyester with the ring in the backbone can be prepared by the Friedel-Crafts polyalkylation of vinyl 2-furoate. $BF_3 \cdot Et_2O$ gave similar results at room temperature in methylene chloride. Structure *30* was arrived at by spectroscopy and by alcoholysis of the polymers.

The only other study of the polymerization of a furanic vinyl ester concerns allyl 2-furylacrylate[120], but the conditions used were so violent (bulk at 200 °C with benzoyl peroxide) that no conclusions meaningful in the present context can be drawn.

4. Furfuryl Acrylate and Furfuryl Methacrylate

The furfuryl esters of acrylic and methacrylic acid polymerize *via* a free-radical mechanism without apparent retardation problems arising from the presence of the furan ring. Early reports on these systems described hard insoluble polymers formed in bulk polymerizations and the cross-linking ability of as little as 2% of furfuryl acrylate in the solution polymerization of methylacrylate[121].

The studies of Mikhailov, Budevska and Berlin[122] showed that solution polymerization of furfuryl methacrylate gave soluble polymers; only when the process was continued to very high conversions were cross-linked materials formed. When the maleic anhydride and maleimide adducts of this monomer[123] were polymerized in benzene with azo-*bis*-isobutyronitrile[124], precipitation of a gel was observed; it was argued that polymerization of both adduct and furfuryl methacrylate (formed in solution by the reverse Diels-Alder reaction) took place and thus the product was probably a copolymer. These authors[125] also studied the course of the bulk polymerizations of furfuryl methacrylate and 5-carbomethoxy-2-furfuryl methacrylate up to high conversions. The yield-*vs.*-time curves were typical for acrylic monomers with self-acceleration due to the considerable increase in viscosity, and sudden paralysis due to the immobilization of the free radicals. The 5-substituted monomer was found to polymerize more rapidly than the unsubstituted one. This observation together with their different behaviour in copolymerizations with styrene[126, 127] were explained on the basis of a different degree of π-complexation between the methacrylic group and the furan ring (much lower in 5-carbomethoxy-2-furfuryl methacrylate because of the electron-accepting character of the 5-substituent). An alternative interpretation of these phenomena could arise from a comparison of a similar difference in behaviour between vinyl 2-furoate and its 5-methyl homologue (see Section III-3), i.e. the presence of a substituent at the C-5 position can reduce the frequency of attack of the furan ring by free radicals. Of course, since with the furfuryl acrylic esters the radical polymerization seemed to be practivally "normal", while with vinyl 2-furoates strong inhibition was observed, any argument about the possible negative role of the furan ring in the former monomers is rather delicate. The values of copolymerization parameters of these monomers with styrene[126, 127] and methyl methacrylate[128] are not very enlightening; they just confirm that the presence of the furan ring does not seem to influence greatly the polymerizability of acrylic and methacrylic monomers.

Recently, Ito and co-workers[129] reported the stereospecific polymerization of 2-furfuryl methacrylate at −78 °C using n-butyllithium, lithium aluminium hydride and isobutyl magnesium bromide. A very high degree of isotacticity was found in the polymer prepared in toluene, while those obtained in tetrahydrofuran were highly syndiotatic, as assessed by NMR signals of the α-methyl protons. This is undoubtedly the first clear report of a highly stereoregular vinyl polymer containing the furan ring.

5. Vinyl Ethers and Vinyl Carbonyl Compounds of the Furan Series

a) Vinyl Ethers. In a short paper dealing with synthetic aspects Rychkova and Keller[130] stated that 2-furfuryl vinyl ether polymerized with $FeCl_3$ to give a viscous liquid. No details of this preparation were given. It is not surprising that a vinyl ether should polymerize with cationic initiators and it is curious that such systems have not received more attention, considering the simplicity of the monomer's synthesis.

Uvarova and Gorshkova[131] reported the thermal polymerization of 2-furfuryl vinyl ether and of a series of vinyl ethers of alkyl-1-(2-furyl) carbinols. Treatment of these compounds at 150 °C for 24 hours produced brown cross-linked products with 2-furfuryl vinyl ether, and soluble, low molecular-weight polymers with the others. It seems likely that in all these systems the ring participated in the growth of the chains, but the lack of experimental details and of structural determination makes it difficult to assess the results.

The divinylacetal of 2-furaldehyde was polymerized *in vacuo* with free-radical initiators at 80 °C to give low yields of the cyclic polyacetal *31*[132], commonly known as poly(vinylfurfural).

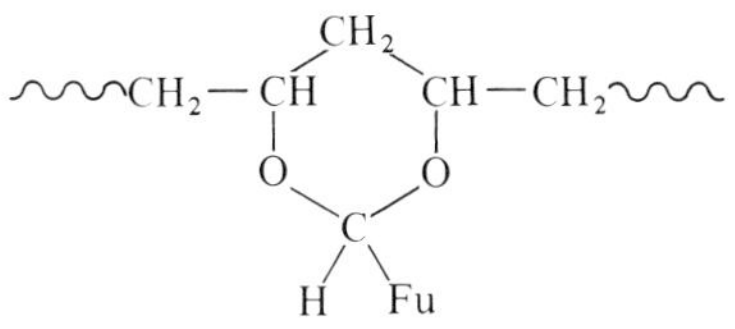

The structure of this product was clearly established. At 150 °C the same process gave cross-linked polymers if the reaction was not stopped before 65% conversion (— the furan ring was probably responsible for side reactions with monomeric vinyl groups). A comparison of the rates of these polymerizations with those obtained using divinylbenzal showed a marked difference in favour of the latter, indicating once again that the presence of the furan ring in a monomer can reduce considerably its radical polymerizability. Also, the molecular weight of the furan polymer was lower than that of its benzene analog. Panayotov, Schopov and Obreschkov[133)] confirmed the reduced tendency of this monomer towards radical polymerization: the highest yield obtained was about 9% for 100 h polymerization at 100 °C with 5% azo-*bis*-isobutyronitrile, *i.e.* a very substantial self-retardation if one also considers that the monomer is divinylic and should, therefore, display high reactivity.

b) Unsaturated Aldehydes and Ketones. 2-Furyl vinyl ketone and 2-furyl isopropenyl ketone were synthesized by Barnes in 1938[134)], but thereafter no interest in these compounds has materialized. It was claimed that both gave linear polymers by "standard procedures"[3)] and if this were true the reasons for the lack of further investigation is intriguing, since very few furan monomers behave in such a straightforward fashion.

In contrast with the above situation, the polymerization of 2-furfurylidene methyl ketone, di-2-furfurylidene ketone and their homologues has been the subject of a large volume of (mainly technical) publications because of the useful applications of the final cross-linked products. As pointed out in the introduction, this review does not deal with the technological aspects of furan resins and in this section only the mechanistic aspects of the first phase of these polymerizations will be discussed.

The well-known condensation between 2-furaldehyde and acetone in a basic medium yields what is usually called "furfurylidene acetone monomer" composed of a mixture of 2-furfurylidene methyl ketone, di-2-furfurylidene ketone, mesityl oxide and other oligomers derived from further condensation reactions[135)]. This mixture is then polymerized by the action of an acidic catalyst; in the first phase of the reaction a polymer of low molecular weight is produced which on further treatment cross-links to a black insoluble and heat-resistant material[136)].

Conflicting interpretations of the mechanism responsible for the acid-catalysed polymerization have been proposed, particularly by the prolific group of Kamenskii[136)] and by Isacescu, Gavat and Ursu[137)]. Essentially, the problem consisted in establishing the predominance of one or more of the following reactions in the initial phase of the process:

— polymerization through the 1,2-disubstituted olefinic bond;
— aldol-crotonic polycondensation between methyl and carbonyl groups to give new unsaturations as the chain grows; this reaction of course would only be possible with 2-furfurylidene methyl ketone;
— successive alkylation of C-5 ring positions;
— ring cleavage and polymerization by the unsaturated compounds produced.

While Kamenskii and co-workers[136)] favoured a mechanism involving reactions of carbonyl, ethylenic and 5-C positions, Isacescu, Gavat and Ursu[137)] proposed that the ring scission was predominant for the polymerization of 2-furfurylidene methyl ketone but not with di-2-furfurylidene ketone. A series of papers by Kamenskii, Usmanov et al.,[138] with a homologous series of ketones did not add any new evidence to this confused situation. It must be pointed out that in all the work cited above poor evidence was given for the mechanistic arguments put forward.

Rodriguez and Gandini[139, 140] have recently carried out some work on the structure of the soluble polymers of the two ketones. The purified monomers were polymerized with various acids to give dark soluble products with DP's of 10–20. The ultraviolet, infrared, and NMR spectra and the elemental analysis of these purified substances were compared with those of the starting monomers. It was concluded that, at least for this initial phase, the two systems are characterized by polymerization through the olefinic bond because:

– the elemental composition of the oligomers is close to that of the monomers (so that condensation reactions cannot be predominant);
– conjugation is lost in the process as shown in the electronic spectra;
– no evidence for significant alkylation at C-5 is obtained from the infrared and NMR spectra;
– the major carbonyl band in the product from 2-furfurylidene methyl ketone appears at 1725 cm^{-1} (monomer 1675 cm^{-1}), indicating loss of conjugation;
– all the typical infrared features of 2-substituted furan rings are maintained.

Of course, these conclusions do not rule out completely the occurrence of other reactions such as those listed above, but their contribution to the overall mechanism must be very small in the production of the oligomers. The dark colour of these products was attributed to hydride transfer reactions, similar in nature to those encountered in the cationic polymerization of 2-vinyl furan [see Section III-B-1-c)]. The subsequent process which transforms these oligomers into cross-linked resins was not investigated.

The polymerization of β-(2-furyl) acrolein and some of its homologues by acids[141] gave a similar phenomenology and thus, presumably, the mechanism is similar to that of the corresponding ketones. Their γ-radiation polymerization only proceeded in solution, probably because acidic substances were formed from the solvent[142].

2-furfurylidene phenyl ketone did not homopolymerize with free-radical initiators. However, it gave low molecular-weight copolymers in poor yields with butadiene, styrene and isoprene, but not with vinyl acetate, acrylonitrile, vinyl chloride, and acrylic monomers[143].

6. Polymerization of 2-Furaldehyde and Homologues through the Carbonyl Bond

This section deals with investigations specifically aimed at producing homopolymers and copolymers of furan carbonyl compounds by the selective opening of the carbonyl bond. The many reports on "polymerization" of 2-furaldehyde which in fact deal with complicated acid-catalysed resinification reactions which involve both the formyl group and the furan ring are reviewed in Chapter VI.

The problem of successful induction of polyaddition through the carbonyl bond in aldehydes and ketones is thermodynamic and does not, therefore, depend on such typical kinetic factors as reaction time or choice of catalyst. It is well known that, in order to overcome an often delicate and unfavourable balance between the decrease in entropy implicit in the process of polymerization and the relatively small values for the enthalpy of the transformation of a C=O bond into a C–O–C bond, it is necessary to carry out the polymerization reaction at low or very low temperatures. Within the realm of aliphatic aldehydes, this situation has been thoroughly studied and satisfactory solutions have been found both regarding the most appropriate conditions for the synthesis of high molecular-weight polyacetals and the more complicated issue of the stabilization of the chain end groups against reactivation of the chain, and subsequent depropagation at room temperature. In the case of ketones, despite many attempts and conflicting reports, particularly with acetone, the situation is still at a difficult stage of development. The same can be said for aromatic and other highly conjugated aldehydes: it has been calculated that the ceiling temperature for the polymerization of benzaldehyde is about $-160\,°C$[144] and no linear polyacetal has been reported to date for this compound. An empirical criterion[145] of the "polymerizability" of carbonyl compounds affirms that, if conjugation provokes the lowering of the infrared absorption frequency below about $1725\,cm^{-1}$, the polymerization through that bond is thermodynamically precluded. This difficulty has been overcome in some instances by the use of special techniques, such as the "Kargin apparatus",[146] which apparently functions by the thermodynamic subterfuge of gaining some entropy by inducing the polymerization at the solid-liquid interphase. Another way of by-passing the thermodynamic restrictions consists in making the carbonyl compound copolymerize with a suitable co-monomer if the enthalpy of the alternate propagation step is higher than that of the restricted carbonyl homopolymerization step. In these situations it is expected that the carbonyl compound will enter the polymer chain but will always be followed by the co-monomer.

2-Furaldehyde is a classical example of such thermodynamically unfavoured "monomers". Its strong conjugation with the ring is well represented by a carbonyl frequency at about $1670\,cm^{-1}$ and the best indication of its reluctance to polymerize is simply the fact that, despite many attempts and some claims of success, no one has in reality been able to prepare a polyacetal with the structure given below:

$$-O-\underset{Fu}{\underset{|}{\overset{H}{\overset{|}{C}}}}-O-\underset{Fu}{\underset{|}{\overset{H}{\overset{|}{C}}}}-O-\underset{Fu}{\underset{|}{\overset{H}{\overset{|}{C}}}}-$$

32

Both anionic[147, 148] and cationic catalysts[148, 149] have failed to promote the homopolymerization of 2-furaldehyde and some of its derivatives. Recently, Gandini and Rieumont[148] also experimented with the Kargin apparatus, but the results were negative.

Copolymerization of 2-furaldehyde through the carbonyl bond has given some encouraging results. Natta and co-workers[147] were the first to describe an alternating copolymer of 2-furaldehyde with dimethyl ketene, obtained at low temperature with anionic initiators. The polyester produced had a stereoregular structure, indicating that alternate addition of the two monomers had taken place. Elemental analysis and chemical breakdown proved this assumption and supporting evidence was given by the sharp melting point of the polymer fraction insoluble in benzene. No molecular weight was given for these products. The structure of this special polyester, prepared by a chain reaction, was:

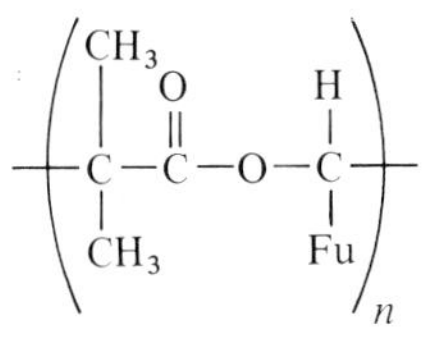

$$\left(\begin{array}{c} CH_3 \\ | \\ -C-C-O-C- \\ | \quad \quad | \\ CH_3 \quad Fu \end{array} \begin{array}{c} O \\ \| \\ \\ H \\ | \\ \end{array} \right)_n$$

33

Kunitake, Yamaguchi and Aso[149] studied the copolymerization of 2-furaldehyde with olefins and vinyl ethers using $BF_3 \cdot Et_2O$ in methylene chloride or toluene at $-78\,°C$. No copolymers were obtained with olefins, but p-tolyl vinyl ether or 2,3-dihydropyran gave polyethers. With the former co-monomer the values of the reactivity ratios were $r_1 = 0.15 \pm 0.15$ and $r_2 = 0.25 \pm 0.05$ (M_1 = 2-furaldehyde). A more complicated behaviour was obtained with divinyl ether due to the formation of both cyclic structures and pendent vinyl groups in the chain. The failure of such olefins as styrene and isopropenylbenzene to give copolymers with 2-furaldehyde, and in fact to homopolymerize in its presence, was blamed on the strength of the complex formed between the initiator and the aldehyde, believed too stable to initiate polymerization.

Gandini and Rieumont[119, 148] conducted a similar parallel study of these systems and obtained at $-78\,°C$ essentially the same results as the preceding authors, but at higher temperatures ($-30-0\,°C$) copolymerization of 2-furaldehyde and 5-methyl-2-furaldehyde with some olefins was observed. A systematic investigation was, therefore, carried out with both aldehydes and a series of co-monomers ranging in basicity from cyclopentadiene to N-vinylcarbazole. The results with 2-furaldehyde can be classified broadly into three categories, according to the qualitative behaviour observed:

– Inhibition of olefin polymerization occurred when its basicity was not sufficient to produce an appreciable displacement of initiator from the aldehyde-acid complex; isoprene, cyclopentadiene and styrene were in this category.

– Copolymerization occurred when the olefin had a basicity lower than that of the aldehyde (with respect to the initiator used), but sufficiently high occasionally to displace a molecule of initiator and give rise to an active species; this situation produced copolymers with varying proportions of ether units in the chain, depending on the monomers feed ratio and on the olefin used. Isopropenylbenzene gave the best results with alternate copolymerization over a fairly wide range of feed ratios: $r_1 = 0.03 \pm 0.03$, $r_2 = 0.4 \pm 0.1$ (2-furaldehyde = M_1); indene produced copolymers with lower 2-furaldehyde contents.

— Homopolymerization of the olefin occurred if its basicity was higher than that of 2-furaldehyde, i.e. the initiator displacement from the complex was strong and thereafter only occasional inclusion of aldehyde in the growing polymer chain took place: 2-furaldehyde units in the products were less than 5%. This situation was observed with acenaphthylene and N-vinylcarbazole.

5-Methyl-2-furaldehyde gave a similar overall behaviour, but a penultimate effect was observed in its copolymerization with isopropenylbenzene whereby two molecules of the aldehyde could add together if the penultimate unit in the growing chain was from the olefin. This was borne out by the copolymers' composition and spectra. The values of the reactivity ratios showed this interesting behaviour: $r_1 = 1.0 \pm 0.1$, $r_2 = 0.0 \pm 0.1$. An apparent paradox occurred: the aldehyde, which could not homopolymerize, had equal probability of homo- and copolymerization and the olefin, which homopolymerized readily, could only alternate. The structure arising from this situation was close to a regular sequence of the type:

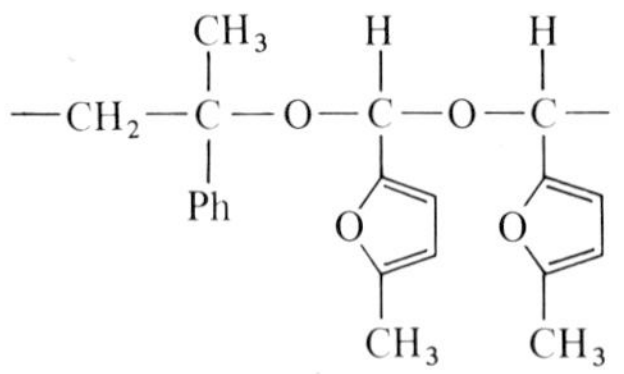

34

2-Furyl methyl ketone was also tried with a series of olefins: inhibition was operative in all systems and only traces of a copolymer with acenaphthylene were isolated at the end of a run which lasted several days. These observations are compatible with the higher basicity of this ketone compared with that of the corresponding aldehyde.

Other aspects of these copolymerizations were studied including some kinetic features, molecular weights (which were low), the extent of branching due to side reactions on the furan ring, and the characterization of the complex formed between 2-furaldehyde and $BF_3 \cdot Et_2O$.

Durgaryan and co-workers[150] reported an investigation of the system styrene 2-furaldehyde in the presence of Lewis acids and dibenzoyl peroxide and claimed copolymerization in both systems. The products of the acidic catalysis were obviously the consequence of extensive resinification of the aldehyde with no indication from their infrared spectra of either ether linkages (1000–1100 cm^{-1}) or styrene units (1500 cm^{-1}). This was to be expected from the working conditions chosen for the experiments, i.e. 50 °C in bulk. The products of the free radical catalysis, on the other hand, must have been low molecular-weight poly(styrene), particularly when the concentration of 2-furaldehyde was high, in view of the retarding effect of the latter in radical polymerization (see Chapter V). The oxygen content of these products and the carbonyl band observed in their infrared spectra must be attributed to the terminal benzoyl groups and to some extent to structures derived from the termination of a chain by a stable furylic radical. The values of the reactivity ratios obtained by these authors for both cationic and free radical systems are, therefore, meaningless.

The "polymerization" of 2-furaldehyde by sodium and sodium naphthalene reported by Kulakov and Kamenskii[151] did not produce structure *32*, but rather some resinous oligomeric materials formed through the interactions of the furan ring with the formyl group.

IV. The Transformation of Poly(vinylalcohol) into Furan Polymers

The use of the hydroxyl groups of poly(vinylalcohol) as reactive sites for the preparation of various "unconventional" polymers is well known and indeed the very synthesis of poly(vinylalcohol) is based on a similar but reverse reaction. This general principle has been applied successfully to the synthesis of some vinyl-type furanic polymers, which cannot be made by classical routes.

The acetalization of up to 50% of the hydroxyl groups of poly(vinylalcohol) with 2-furaldehyde in the presence of acidic catalysts has been reported[152, 153]. The polymer was successively polyadducted with maleic anhydride[154, 155]. This so called poly(vinylfurfural) bears in its structure the same vinyldiacetalic unit as product *31*, obtained in the direct polymerization of the divinylacetal of 2-furaldehyde, but with the important differences that it has a much higher molecular weight (as high as that of the starting material) and it is prepared in good yields without problems of autoretardation.

Tsuda[156] reported the preparation of poly(vinyl 2-furylacrylate) by the reaction of 2-furylacrylyl chloride with poly(vinylalcohol) in NaOH-water-methyl ethyl ketone. Up to 80% of the hydroxyl groups were esterified. The interest of this technique is obvious here, considering that the vinyl ester of 2-furylacrylic acid does not polymerize[119]. A similar procedure was employed by Gandini and Rieumont[26, 119] for the synthesis of poly(vinyl 2-furoate) another product unobtainable *via* a standard polymerization process (see Section III-B-3).

The potential of these transformations has not been fully exploited. The method is certainly valuable both for the purpose of studying the structure of furan polymers which might be of interest but cannot be prepared easily by conventional techniques and for preparing polymers in which a certain proportion of the hydroxyl groups of poly(vinylalcohol) are substituted with a furan moiety which gives the new product some special property.

V. The Retarding and Inhitibing Role of Furan Derivatives in Radical Polymerization

In the preceding chapters frequent mention was made of
  – the retarding or inhibiting effect of furan derivatives on the radical polymerization of other monomers and

— the autoinhibition and autoretardation of monomers containing the furan ring when they were submitted to free-radical initiation.

These phenomena will below be discussed more systematically.

A. Strong Generalized Inhibition

Certain furan derivatives can be classified as good inhibitors in the classical meaning of the term, *i.e.* they fulfill the same role as compounds used commercially to stop radical chain reactions. It must be emphasized here that the compounds which belong to this class act indiscriminately in the sense that their capacity of inhibiting a radical polymerization extends to all monomers. Borrows, Haward, Porges and Street[157] discovered that 2-furfurylidenemalonitrile, ethyl 2-furfurylidenecyanoacetate and diethyl 2-furfurylidenemalonate were powerful inhibitors of polymerization. It was suggested that these unsaturated compounds acted by adding onto the growing radical to give a stabilized radical with a long lifetime. Recently, Hernandez, Galego, Llerandi and Gandini[158] reported that furan derivatives with the structure

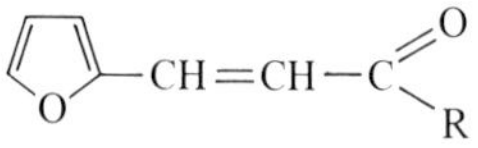

35

where R = H, OH, CH_3, OAlk, Furfurylidene, exhibited similar properties. They argued that the reaction of a growing radical with the olefinic bond could give rise to a resonance stabilized radical of enhanced stability. The radical polymerization of styrene, vinyl acetate, methylmethacrylate and other common monomers was inhibited by the presence of very small concentrations of these compounds.

All the furan derivatives mentioned above possess a degree of conjugation that goes well beyond that of simple furans and their efficiency must be viewed as the result of the composite effect of the ring (see below) plus the conjugated substituents. However, the fact that the corresponding benzene derivatives did not show any inhibiting power and instead copolymerized with styrene[157], shows that the furan ring plays a determining role in promoting the inhibition.

B. Selective Retardation

There are several reports scattered in the literature of the retarding effect of simple furan derivatives in the polymerization of a specific monomer. Hardy[69, 116] found that furan, 2-furoic acid and its esters, and 5-substituted-2-furoic acids were strong retarders in the radical polymerization of vinyl acetate, but did not act likewise with styrene. He proposed that as a result of the reactions of the free radicals with the furan derivatives, dihydro- and tetrahydrofurans would form, but he did not produce any evidence to support these speculations. Clarke, Howard and Stockmayer[159] reported the retarding action of 2-furaldehyde on the radical polymerization of vinyl acetate in a massive study where many compounds were used as potential chain transfer agents. 2-Furaldehyde was one of the strongest "degradative transfer agent" encountered. Savranskaya, Trubitsyna and Askarov[160] mentioned that 2-methylfuran and 2-furaldehyde retard the radical polymerization of acrylonitrile. Similar observations have been quoted in earlier chapters when radical copolymerizations of typical vinyl monomers with furan monomers were discussed.

Gandini and Rieumont[26, 119, 161] investigated this field from two different angles. They studied the reaction of a typical radical initiator, azo-bis-isobutyronitrile, with various furan derivatives, including furan, 2-methylfuran, ethyl 2-furoate and 2-furaldehyde. Addition to the ring predominated in all reactions with formation of dihydrofuran derivatives. Separation of these products was difficult because of the presence of three types of positional isomers and the stereoisomers of each of the latter. With furan, satisfactory separation was achieved and the following compounds characterized:

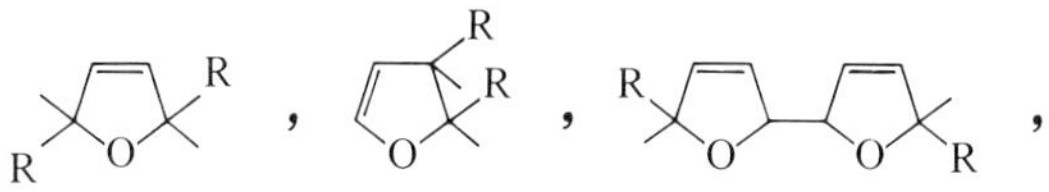

36, two stereoisomers

37, two stereoisomers

38 , and other dimers.

(R = 2-cyano-2-propyl)

With 2-methylfuran, ethyl 2-furoate and 2-furaldehyde similar results were obtained and in all reactions the 2,5-dihydrofuran products were favoured. The mechanism proposed involved the following steps:

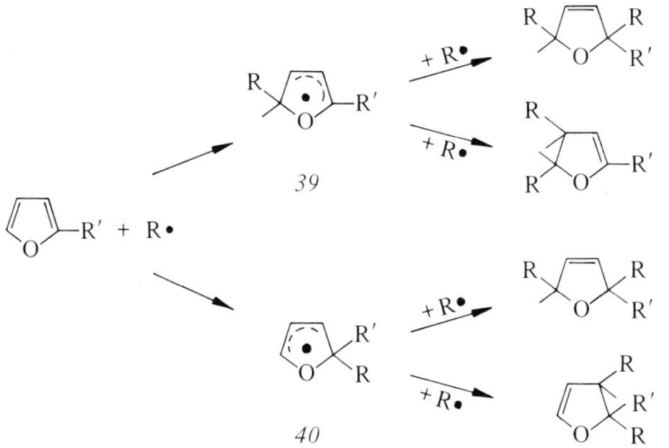

39

40

+ various dimerization steps of 39 and 40.

The absence of higher oligomers indicated that the furylic radicals 39 and 40 were unable to propagate and only coupled with either a primary radical or between themselves. The reaction of this initiator with 2-vinylfuran and vinyl 2-furoate was also studied. The former gave oligomers with a normal vinyl structure without any evidence of dihydrofuranic rings. The latter also gave oligomers, but, as mentioned in Section III-B-3, both furanic and dihydrofuranic rings were present in their structures which also contained vinyl groups. In the second part of this study the relative retarding effect of ethyl 2-furoate on the bulk radical polymerization of methylmethacrylate, styrene and vinyl acetate was assessed. It was found that vinyl acetate was the most affected (inhibition), followed by styrene (retardation with fairly high concentration of furan derivative), while methyl methacrylate polymerized normally in presence of up to 25% of ethyl 2-furoate. Similar results were obtained when 2-methylfuran was used as retarder.

The above results indicate that furan derivatives operate as retarders in radical polymerization by reacting with a growing chain through the C-2 or C-5 position of the ring to form a more stable furylic radical. Retardation will be the more effective, the less stabilized the monomer's radical, *i.e.* the degree of rate and molecular-weight reduction achieved by the furan retarder will depend upon the comparative stability of the growing radical and the furylic radical which can form by the addition of the former to the ring.

C. Autoinhibition

The fact that many monomers containing the furan structure are reluctant to polymerize with free-radical initiators (see Chapter III) is obviously a direct consequence of the above argument. These substances have a built-in retarding substituent and their possibility of forming polymers by a normal polyaddition growth through the vinyl group is hampered. Again, the degree of autoretardation will depend on the relative stability of the two radical moieties that these molecules can produce upon addition to a free radical. Thus, vinyl 2-furoate is strongly self-retarded because the radical formed on the ring is much more stable than that involving the vinyl ester group(cf. furans in vinyl acetate polymerization). At the other limit, the 2-furfuryl acrylates polymerize freely because the free radical formed on the acrylic moiety is more stabilized than that which would form by addition to the furan ring (cf. lack of effect of furan derivatives on the polymerization of methyl methacrylate). Other monomers of the furan series will behave according to this competitive criterion.

2-Vinylfuran represents a special situation. Although its radical polymerization is particularly slow and the molecular weights are low, the structure of these polymers does not reveal any anomaly, even when very high concentrations of initiator are used. Thus, both primary and oligomeric radicals, the latter being rather stable, attack the vinyl bond only and propagation proceeds in a conventional fashion, at least during the first period of the polymerization. Already at moderate yields the rate of polymerization decreases considerably. This behaviour is typical in bulk and solution. In emulsion[85] none of these peculiarities are observed and high yields of high molecular-weight polymers are obtained with reasonable rates. The radical responsible for propagation should be considered as the resonance hybrid resulting from the three canonical structures:

Structures 41, 42, 43

In analogy with the furanylmethyl radical[162], structure *41* is probably the most favoured. In fact, further monomer addition proceeds from it. Moreover, the monomer is attacked at the vinyl group to regenerate the same radical and its ring is not a competitive site for this attack. The absence of detectable amounts of dihydrofuranic structures in the polymers suggests the above conclusions. However, the reasons for such selectivity both in the nature of the propagating species and in its mode of addition to the monomer are not clear. The low rate of propagation must be associated with the stability of the growing species, and the low molecular weights could be due to the unfavourable value of k_p^2/k_t [86]. The autoretardation, which develops after a certain amount of polymer has accumulated in solution, suggests that the unconjugated rings of the polymer can act as alternative sites for the attack of the growing radical, a property not displayed by their conjugated counterparts in the monomer. The fact that these problems do not arise in emulsion could be explained by considering that as long as only one radical is present in the micelle, the chain can continue growing (fairly high molecular weights) and that the concentration of the polymer relative to that of the monomer around an active species is always small (high yields).

All the above conclusion should of course be considered as tentative and a more thorough interpretation of this phenomenology must await further and more specific work.

VI. Stability and Resinification of Furan Derivatives

It is a fairly common feature for furan polymers to suffer cross-linking and colour-forming reactions during their synthesis or after their isolation. Examples of systems which involve a first phase of linear growth followed by a second phase in which less reactive sites promote branching and eventually gelation have been given throughout this review. In other systems the products discolour and become insoluble shortly after isolation and exposure to the atmosphere. While in certain instances these phenomena can be looked upon as positive features in the sense that the specific applications for which the products are intended require cross-linked materials with good resistance to strong chemicals and high temperatures, it is obvious that if the processing of the polymer requires thermoplasticity, then the reactions leading to cross-linking should be suppressed. Likewise, while the colour of some of these products does not affect the requirements for some of their applications, in many situations the progressive darkening of initially colourless furan polymers can be a serious drawback against other potential uses.

The origin of these complications, not common in other polymers, at least on such a short time scale, is connected to the presence of the furan ring or of structures derived from it in the macromolecules. Many furan derivatives also display a lack of stability which is not shared by their aliphatic and aromatic counterparts. The mechanisms of formation of dark resinous materials from these compounds can be studied and subsequently used as models for the understanding of the more complex situations which occur in furan polymers. A brief survey of these topics is given in this chapter.

In connection with their stability, furan derivatives can be broadly classified into threee major categories:

— Compounds which are stable by any normal chemical standard. 2-Furoic acid and its esters are the best example of such normal behaviour, due to the presence of electron-withdrawing substituents which stabilize the ring against acid-catalyzed resinification[163] and oxidation[164].

— Compounds which are rather unstable. Typical members of this class are 2-furaldehyde, 2-furfuryl alcohol and 2-alkyl furans, the latter being more resistant than the former. The action of acids or oxygen on these derivatives produces appreciable resinification, but, if properly purified and stored *in vacuo*, they are indefinitely stable[25, 165].

— Compounds which are highly unstable. 2-Furfuryl chloride[18], 2-furylacetonitrile[18] and difuryl carbinols[24, 28, 29, 166] resinify spontaneously upon isolation and must be kept in solution at low temperature.

Depending on whether the degrading agent is oxygen or an acid, it can be assumed that free radicals or carbenium ions are formed in these processes. The aggregation reactions which ultimately lead to tarry materials proceed through these intermediates and involve both the furan ring and its substituents. The participation of the ring is at least threefold:

— free C-5 positions are vulnerable to addition and substitution reactions;
— electronic rearrangements can occur with formation of dihydrofuranic or similar structures and
— ring cleavage is possible particularly in oxidative degradations or if aqueous acid solutions are used.

The participation of the substituents in the resinification reactions reflects the effect that the ring has on them, and very often the reactivity of a given functionality is substantially enhanced by the neighbouring influence of the furyl moiety.

These general considerations are illustrated below by a few examples.

The behaviour of 2-furaldehyde is informative in many respects. Its sensitivity to acids arises from the participation of both the C-5 positions and the carbonyl groups in initial condensation reactions involving a protonated molecule and its neutral counterparts. It is likely that in these first stages 2-furaldehyde is converted into trifurylic species which can easily form carbenium ions that grow and isomerize to give highly conjugated structures similar to that shown as 5, but with a delocalized positive charge instead of an unpaired electron. This general mechanism is based on recent work[24] and differs somewhat from that proposed by Nakamura and Saito[167] who claimed that ring scission was the major cause for polyconjugation. The observation that neither 5-methyl-2-furaldehyde nor 2-furyl methyl ketone undergo acid-catalyzed resinification[24, 168] supports the former mechanism in that with the first compound condensation is precluded by the C-5 substitution and in the second the methyl group minimizes the formation of trifurylic condensates because of steric hindrance. Of course, hydrolytic ring cleavage can take place if aqueous acid solutions are used, but this reaction does not give rise to dark tarry products.

Furyl carbinols display a tendency to give resinous materials which is the more pronounced the more furyl groups are attached to the same carbon atom. Thus, 2-furfuryl alcohol and its alkyl homologues are more stable than bis(2-furyl) carbinols which have been prepared but not fully characterized because they decompose rapidly[28, 29]; tris(2-furyl) carbinol has never been isolated, but evidence for its existence as a short-lived species was obtained by Gandini and Galego[24] after a synthesis involving butyl lithium and furoyl chloride. Spectra of the ethereal solution after the usual work-up procedure indicated that the carbinol had been formed, but its resinification, accompanied by the appearance of an intense green colour, was rapid even in solution. Pennanen[164] prepared tris(5-methoxycarbonyl-2-furyl) carbinol and showed that the presence of three electron-withdrawing substituents which also blocked the C-5 positions successfully stopped any tendency towards resinification even in concentrated acids. The instability of the carbinols must be related to the ease with which carbenium ions and radicals, stabilized by the adjacent rings, are formed even in mild conditions. Similarly, the spontaneous resinification of 2-furfuryl chloride and 2-furylacetonitrile[18] involves the furfurylenium ion *1* as the initial chain carrier.

The effect of acids on 2-alkylfurans was discussed in Section III-A-1-c). It was remarked that 2,5-disubstituted furans are less reactive in this context than monosubstituted derivatives. This observation again points to the important role played by a free C-5 position in resinification processes.

All the derivatives examined above owe their instability to the presence of the furan ring; similar aldehydes, carbinols, chlorides, etc. bearing aliphatic and aromatic substituents are not prone to resinify. An analogous singularity of behaviour is encountered in furan polymers, as underlined at the beginning of this chapter. In a polymer prepared from a furan derivative three different situations must be considered:

— The ring is still present in the macromolecule and appears as a pendant Fu group. The side reactions, which can alter the normal structure of these macromolecules and ultimately lead to darkening and cross-linking of the product, can originate from the activation of the C-5 position through transfer reactions with polymer during the synthesis or through interchain reactions after isolation of the polymer and from the enhanced reactivity of other groups due to the effect of the neighbouring ring. These reactions were observed in the cationic polymerization of 2-vinylfuran and it was clearly shown [see Section III-B-1-c)] that, by substituting the C-5 and α-hydrogens with methyl groups, a linear colourless stable polymer could be prepared. The gelation of poly(2-vinylfuran) by atmospheric agents must be related to similar weaknesses in its macromolecules, and particularly to the ease of peroxide formation from the tertiary carbon atom in the chain[169]. The other clear example of the occurrence of branching through the C-5 position is the copolymerization of 2-furaldehyde with olefins[148]. Viscosimetric measurements showed that these copolymers were branched, while similar products obtained with 5-methyl-2-furaldehyde were linear. In these systems cross-linking was not attained because the molecular weights were low.

— The ring is in the chain backbone as a 2,5-disubstituted moiety. In these situations the polymer is more resistant to side reactions but some substituents are more susceptible than others to activation by the presence of the ring and promote branching and degradation under the influence of air, particularly at high temperature. This particular behaviour is encountered in some polycondensates as discussed in Section II-A.

— The ring has lost its furan character and is present in the chain backbone as dihydro- or tetrahydrofuranic units. The susceptibility of these polymers to degradation and branching reactions arises from the residual double bond of the dihydrofuranic rings, which is a vinyl-ether function, or from the hydrolytic scission of the tetrahydrofuran rings. Both oxygen and acids can damage these structures. The polymers derived from furan, 2-alkylfurans and dihydrofurans are examples of this situation, as discussed in Sections III-A-1-c) and III-A-4.

The conclusions which can be drawn from this general analysis are that if the C-5 position of the ring is protected, and better still protected by an electron-withdrawing group, polymers containing furan rings will certainly be much more stable. Also, the use of standard antioxidants will reduce the rate of peroxide formation and thus the likelihood of degradation. However, these criteria are not proposed as a general solution to the problems discussed above. A better knowledge of the reactions promoting resinification of simple furans will undoubtedly provide more specific and effective ways of improving the stability of furan polymers.

VII. Conclusions and Acknowledgments

Within the broad spectrum of systems described, only few can be considered well understood. Unfortunately, many past investigations were superficial and more involved studies have been avoided because of the inherent complex phenomenology. This has been harmful, and misleading conclusions have been reached about the real potential of furan monomers. It is hoped that this review has shown the fallacy of this approach and will therefore stimulate further skilled work in this field.

I am deeply grateful to my former colleagues Rubén Alvarez, Heriberto Campañá, Norma Galego, Carlos Hernandez, Ricardo Martínez, Jacques Rieumont, and Silvia Prieto for their valuable contributions to the knowledge of furan chemistry and for making my long stay in Cuba such a fulfilling experience.

VIII. References

1) Cook, M. J., Katritzky, A. R., Linda, P.: Adv. Heterocyclic Chem. **17**, 255 (1974)
2) Albert, A.: Heterocyclic chemistry. 2nd edit. New York: Oxford University Press, p. 278
3) Dunlop, A. P., Peters, F. N.: The furans. New York: Reinhold Publ. Co. 1953
4) Heertjes, P. M., Kok, G. J.: Delft Progr. Rep. **A1**, 59 (1974)
5) Kelly, J. E.: Ph. D. Thesis, Rensselaer Polytechnic Inst. 1975. Xerox Univ. Microfilms 76–3698
6) Ogata, N., Shimamura, K.: Polymer J. **7**, 72 (1975); Polymer Preprints ACS **17**, 151 (1976)
7) Primelles, E.: Ph. D. Thesis, Universidal Central, Santa Clara, Cuba 1977
8) Schmitt, C. R.: Polymer-plastics technology and enerineering. Vol. 3. New York: Marcel Dekker 1974, p. 121
9) Hachihama, Y., Shono, T.: Technol. Rep. Osaka Univ. **7**, 479 (1957)
10) Barr, J. B., Wallon, S. B.: J. Appl. Polymer Sci. **15**, 1079 (1971)
11a) Wewerka, E. M., Loughran, E. D., Walters, K. L.: J. Appl. Polymer Sci. **15**, 1437 (1971)
11b) Wewerka, E. M.: J. Polymer Sci. A**19**, 2703 (1971)
12) Hachihama, Y., Shono, T.: Technol. Rep. Osaka Univ. **4**, 413 (1954)
13) Rathi, A. K. A., Chanda, M.: J. Appl. Polymer Sci. **18**, 1541 (1974); Krishnan, T. A., Chanda, M.: Angew. Makromol. Chem. **43**, 145 (1975)
14) Conley, R. T., Metil, I.: J. Appl. Polymer Sci. **7**, 37 (1963)
15) Conley, R. T., Metil, I.: J. Appl. Polymer Sci. **7**, 1083 (1963)
16) Marino, G.: Adv. Heterocyclic Chem. **13**, 235 (1971)
17) Politzer, P., Weinstein, H.: Tetrahedron **31**, 915 (1975)
18) Divald, S., Chun, M. C., Joullié, M.: J. Org. Chem. **41**, 2835 (1976)
19) Nakamura, Y., Saito, M.: Kogyo Kagaku Zasshi **62**, 1173 (1959); C.A. **57**, 13985 (1962)
20) Marcusson, J.: Z. Angew. Chem. **32**, 113 (1919)
21) Dunlop, A. P., Peters, F. N.: Ind. Engng. Chem. **32**, 1639 (1940)
22) Illari, G.: Gazz. Chim. Ital. **77**, 389 (1947)
23) Nakamura, Y., Saito, M.: Kogyo Kagaku Zasshi **62**, 1179 (1959); C.A. **57**, 8523 (1962)
24) Galego, N.: Ph. D. Thesis, University of Havana, 1975
25) Galego, N.: Gandini, A.: Revista CENIC, Sci. Fis., **6**, 163 (1975); C.A. **84**, 18211 (1976)
26) Gandini, A., Rieumont, J.: Tetrahedron Letters **1976**, 2101
27) Gilman, H., Breuer, F.: J. Am. Chem. Soc. **56**, 1123 (1934)
28) Ramanathan, V., Levine, R.: J. Org. Chem. **27**, 1216 (1962)
29) Heathcock, C. H., Gulick, L. G., Dehlinger, T.: J. Heteroc. Chem. **6**, 141 (1969)
30) Normant, H., Angelo, B.: Bull. Soc. Chim. (France) **1961**, 1988
31) Goutière, G., Golé, J.: Bull. Soc. Chim. (France) **1965**, 162
32) Kasai, P. H., McLeod, Jr., D.: J. Am. Chem. Soc. **95**, 4801 (1973)
33) Armour, M., Davies, A. G., Upadhyay, J., Wassermann, A.: J. Polymer Sci. A1 **5**, 1527 (1967)
34) Kresta, J., Livingston, H. K.: Polymer Letters **8**, 795 (1970)
35) Livingston, H. K., Senkus, R., Tai-Tung Hsien, J., Kresta, J., Makromol. Chem. **161**, 101 (1972)
36) Hachihama, Y., Shono, T., Ishigaki, A.: Technol. Rep. Osaka Univ. **13**, 481 (1963)
37) Korshak, V. V., Sultanov, A. S., Abduvaliyev, A. A.: Uzb. Khim. Zh. **1959** (4), 39; C.A. **54**, 12642 (1960)
38) Topchiev, A. V., Goldfarb, Y. Y., Krentsel, B. A.: Vysokomol. Soedin. **3**, 870 (1961)
39) Khaidarov, K. F., Abduvaliyev, A. A., Sultanov, A. S.: Vysokomol. Soedin. **5**, 1012 (1963)
40) Khaidarov, K. F., Abduvaliyev, A. A., Sultanov, A. S.: Uzb. Khim. Zh. **1964** (4), 65; C.A. **62**, 6683 (1965)
41) Ishigaki, A., Shono, T., Hachihama, Y.: Kogyo Kagaku Zasshi **66**, 119 (1963); C.A. **59**, 4047 (1963)
42) Ishigaki, A., Shono, T.: Bull. Chem. Soc. Japan **47**, 1467 (1974)
43) Granzov, A., Wendenburg, J., Henglein, A.: Z. Naturforsch. **B19**, 1015 (1964)

44) Livingston, H. K., Senkus, R.: Polymer Letters 7, 635 (1969)
45) Topchiev, A. V., Krenstel, B. A., Goldfarb, Y. Y.: Izv. Akad. Nauk. SSR, Otd. Khim. Nauk 1969, 369
46) Gandini, A., Hernandez, C.: unpublished results
47) Murahashi, S., Nozakura, S., Hatada, K.: Bull. Chem. Soc. Japan 34, 939 (1961)
48) Karasev, V. N., Minsker, K. A.: Vysokomol. Soedin. A 15, 1266 (1973)
49) Boucheron, B., Froelich, H.: Ger. Pat. 2, 345, 515 (1974)
50) Butler, G. B., Badgett, J. T., Sharabash, M.: J. Macromol. Sci.-Chem. A 4, 51 (1970)
51) Gaylord, N. G., Maiti, S., Patnaik, B. K., Takahashi, A.: J. Macromol. Sci.-Chem. A 6, 1459 (1972)
52) Gaylord, N. G., Maiti, S.: J. Macromol. Sci.-Chem. A 6, 1481 (1972)
53) Kamo, B., Morita, I., Horie, S., Furusawa, S.: Polymer J. 6, 121 (1974)
54) Vazzana, I., Grandi, F., Hayashi, K., Munari, S., Russo, S.: Chim. Ind. (Milan) 57, 745 (1975)
55) Ragab, Y. A., Butler, G. B.: J. Polymer Sci., Polymer Letter Ed. 14, 273 (1976)
56) Kraemer, G., Spilker, A.: Ber. 23, 78 (1890)
57) Stoermer, R.: Ann. 312, 237 (1900)
58) Sigwalt, P.: Compt. Rend. 252, 3800 (1961)
59) Mizote, A., Tanaka, T., Higashimura, T., Okamura, S.: J. Polymer Sci. A1 4, 869 (1966)
60) Natta, G., Farina, M., Peraldo, M., Bressan, G.: Chim. Ind. (Milan) 43, 161 (1961)
61) Natta, G., Farina, M., Peraldo, M., Bressan, G.: Makromol. Chem. 43, 68 (1961)
62) Farina, M., Bressan, G.: Makromol. Chem. 61, 79 (1963)
63) Natta, G., Bressan, G., Farina, M.: Rend. Accad. Naz. Lincei Sc. Fis. Mat. Nat. 34, 475 (1963)
64) Panton, C. J., Ph. D. Thesis, Birmingham, 1963. See also Plesch, P. H. (ed.): The chemistry of cationic polymerization. Oxford: Pergamon Press 1963, p. 450
65) Takeda, Y., Hayakawa, Y., Fueno, T., Furukawa, J.: Makromol. Chem. 83, 234 (1965)
66) Bressan, G., Broggi, R.: Chim. Ind. (Milan) 50, 1326 (1968)
67) Bressan, G.: Chim. Ind. (Milan) 51, 705 (1969)
68) Bressan, G., Farina, M., Natta, G.: Makromol. Chem. 93, 283 (1966)
69) Hardy, G., Nyitrai, K.: IUPAC Symposium on Macromolecular Chemistry, Wiesbaden, 1959, Preprint III B 6
70) Kamo, B., Kurashige, S., Furusawa, S.: Kogyo Kagaku Zasshi 73, 580 (1970); C.A. 73, 46154 (1970)
71) Barr D. A., Rose, J. B.: J. Chem. Soc. 1954, 3766
72) Schmitt, G. J., Schuerch, C.: J. Polymer Sci. 49, 287 (1961)
73) Kamo, B., Suzuki, N. Ogawa, K. Furusawa, S.: Nippon Kagaku Kaishi 1973, 610; C.A. 79, 5657 (1973)
74) Minoura, Y., Mitoh, M.: Makromol. Chem. 119, 104 (1968)
75) Gandini, A., Galego, N.: J. Polymer Sci. Polymer Chem. Ed. 15, 1027 (1977)
76) Gandini, A., Parsons, J., Back, R. A.: Can. J. Chem. 54, 3095 (1976)
77) Andrews, D. J., Feast, W. J.: J. Polymer Sci., Polymer Chem. Ed. 14, 331 (1976)
78) Berlin, A. A., Liogonski, B. I., Zapadinski, B. I., Kazanzeva, E. A., Stankevitch, A. O.: Vysokomol. Soedin. 18 A, 926 (1976)
79) Izumi, T., Iino, T., Kasahara, A.: Bull. Chem. Soc. Japan 46, 2251 (1973)
80) Lohmann, Y.: Thesis, Nancy University, 1969
81) Leonard, E. C. (ed.): Vinyl and diene monomers. Part. 3. New York: Wiley Interscience 1971, p. 1529
82) Aso, C., Kunitake, T., Tanaka, Y., Miyasaki, H.: Kobunshi Kagaku 24, 187 (1967); C.A. 68, 22275 (1968)
83) Aso, C., Kunitake, T., Tanaka, Y.: Bull. Chem. Soc. Japan 38, 675 (1965)
84) Koton, M. M.: J. Polymer Sci. 30, 331 (1958)
85) Simek, I., Hanus, M.: Chem. Zvesti 14, 124 (1960)
86) Aso, C., Tanaka, Y.: Kobunshi Kagaku 21, 373 (1964); C.A. 62, 9239 (1965)
87) Kamenar, S., Simek, I., Regensbogenova, E.: Chem. Zvesti 14, 581 (1960)

88) Reikhsfeld, V. O.: Tien Tsin Ta Hsueh Hsueh Pao **45**, 200 (1957); C.A. **56**, 7475 (1962)
89) Trifonov, A., Panayotov, I.: Compt. Rend. Acad. Bulg. Sci. **10**, 301 (1957)
90) Alvarez, R., Martinez, R.: M. Sc. Thesis, University of Havana, 1971
91) Trifonov, A., Panayotov, I.: Compt. Rend. Acad. Bulg. Sci. **10**, 353 (1957)
92) Hiraoka, H.: J. Phys. Chem. **74**, 574 (1970)
93) Martinez, R.: Ph. D. Thesis, University of Havana, 1975
94) Alvarez, R., Gandini, A., Martinez, R.: J. Polymer Sci., Polymer Letters Ed. **13**, 385 (1975)
95) Hernandez, C., Gandini, A.: unpublished results
96) Bachman, G. B., Heisey, L. V.: J. Am. Chem. Soc. **71**, 1985 (1949)
97) Polyakova, A. M., Korshak, V. V., Lipatnikov, N. A.: Vysokomol. Soedin. **4**, 334 (1962)
98) Bachman, G. B., Filar, L. J., Finholt, R. W., Heisey, L. V.: Ind. Eng. Chem. **43**, 997 (1951)
99) Andreeva, I. V., Koton, M. M.: Doklady Akad. Nauk SSR **110**, 75 (1956)
100) Goutière, G., Léonetti, J. B., Golé, J.: Compt. Rend. **257**, 2485 (1963)
101) Kuwata, K., Kawazura, H., Hirota, K.: Nippon Kagaku Zasshi **81**, 1770 (1960); C.A. **56**, 4928 (1962)
102) Hernandez, C.: Ph. D. Thesis, University of Havana, 1977
103) Stoicescu, C., Dimonie, M.: Rev. Roum. Chim. **13**, 109 (1968)
104) Gandini, A., Martinez, R., Sanchez, R.: unpublished results
105) Gandini, A., Martinez, R.: J. Polymer Sci., Symposia, in press
106) Alvarez, R., Gandini, A., Martinez, R.: submitted to Makromol. Chem.
107) Gandini, A., Martinez, R.: submitted to Makromol. Chem.
108) Plesch, P. H.: Progress in high polymers. Vol. 2, Robb, I.C. (ed.). London: Iliffe Books, 1968, p. 158
109) Gandini, A., Plesch, P. H.: J. Chem. Soc. **1965**, 6019
110) Mishina, A.: Nippon Nogeikagaku Kaishi **34**, 649 (1960); C.A. **58**, 14109 (1963)
111) Wiley, R. H., Smith, N. R.: J. Am. Chem. Soc. **72**, 5198 (1950)
112) Adrova, N. A., Koton, M. M.: Vysokomol. Soedin. **2**, 408 (1960)
113) Topchiev, A. V., Alaniya, V. P.: J. Polymer Sci. **A, 1**, 599 (1963)
114) Hopff, H., Lussi, H.: Makromol. Chem. **18/19**, 227 (1956)
115) Hardy, G., Szita, J.: Acta Chim. Acad. Sci. Hung. **15**, 338 (1958)
116) Hardy, G.: Acta Chim. Acad. Sci. Hung. **17**, 121 (1958)
117) Spektor, V. I.: Izv. Akad. Nauk Moldavsk. SSR **1966** (9), 65; C.A. **69**, 77831 (1968)
118) Morita, I., Kato, Y., Kamo, B., Furusawa, S.: Chuo Kaygaku Rikag. Kiyo **16**, 139 (1973); C.A. **81**, 106028 (1974)
119) Rieumont, J.: Ph. D. Thesis, University of Havana, 1975
120) Trifonov, A., Panayotow, I.: Acta Chim. Acad. Sci. Hung. **18**, 487 (1959)
121) Rehberg, C. E., Fisher, C. H.: J. Org. Chem. **12**, 226 (1947)
122) Mikhailov, M., Budevska, K., Berlin, A. A.: Compt. Rend. Acad. Bulg. Sci. **19**, 807 (1966)
123) Mikhailov, M., Budevska, K., Berlin, A. A.: Compt. Rend. Acad. Bulg. Sci. **19**, 909 (1966)
124) Mikhailov, M., Budevska, K., Berlin, A. A.: Compt. Rend. Acad. Bulg. Sci. **19**, 1019 (1966)
125) Mikhailov, M., Budevska, K.: Makromol. Chem. **117**, 80 (1968)
126) Berlin, A. A., Budevska, K., Mikhailov, M.: Izv. Akad. Nauk SSSR, Ser. Khim. **1966**, 943
127) Budevska, K., Berlin, A. A., Mikhailov, M.: Vysokomol. Soedin. **9B**, 309 (1967)
128) Bevington, J. C., Harris, D. O.: Polymer Letters **5**, 799 (1967)
129) Ito, T., Aoshima, K., Toda, F., Uno, K., Iwakura, Y.: Polymer J. **1**, 278 (1970)
130) Rychkova, A. G., Keller, R. E.: Zh. Obsch. Khim. **31**, 1849 (1961)
131) Uvarova, N. I., Gorschkova, P. R.: Izv. Sib. Otd. Akad. Nauk. SSR, Ser. Khim. Nauk, **1964** (1), 41; C.A. **61**, 11950 (1964)
132) Matsoyan, S. G., Akopyan, L. M.: Vysokomol. Soedin. **3**, 1311 (1961)
133) Panayotov, I. M., Schopov, I., Obreschkov, A.: Makromol. Chem. **100**, 41 (1967)
134) Barnes, C. E.: U.S. Patent **2**, 309, 727 (1943)
135) Isacescu, D. A., Gavat, I., Stoicescu, C., Vass, C., Petrus, I.: Rev. Roum. Chim. **10**, 219 (1965)
136) Kamenskii, I. V., Ungarean, N. V., Kovarskaya, B. M., Itinski, V. I.: Plast. Massy **1960** (12), 9

137) Isacescu, D. A., Gavat, I., Ursu, V.: Rev. Roum. Chim. **10,** 257 (1965)
138) Usmanov, Z., Kamenskii, I. V., Losev, I. P.: Uzbek. Khim. Zh. **1965** (1), 47, and references therein. Also, Usmanov, Z., Askarov, M. A.: Uzbek. Khim. Zh. **1969** (3), 38; C.A. **63,** 1941 (1965) and C.A. **71,** 92187 (1969)
139) Rodriguez, V. J., Gandini, A.: Revista CENIC, Sci. Fis. **5,** 29 (1974); C.A. **82,** 58412 (1975)
140) Rodriguez, V. J., Gandini, A.: Revista CENIC, Sci. Fis. **6,** 155 (1975); C.A. **84,** 5481 (1976)
141) Usmanov, Z., Askarov, M. A., Kamenskii, I. V.: Uzbek, Khim. Zh. **1969** (1), 50, and references therein; C.A. **70,** 115661 (1969)
142) Yamakita, H., Hayakawa, K.: J. Polymer Sci. Al, **5,** 3219 (1967)
143) Marvel, C. S., Peterson, W. R., Inskip, H. K., McCorkle, J. E., Taft, W. K., Labbe, B. G.: Ind. Engng. Chem. **45,** 1532 (1953)
144) Aso, C., Tagami, S., Kunitake, T.: J. Polymer Sci. Al **7,** 497 (1969)
145) Pregaglia, G. F., Binaghi, M.: The stereochemistry of macromolecules. Ketley, A. D. (ed.). New York: Marcel Dekker, 1967, p. 114
146) Kargin, V. A., Kabanov, V. A., Zubov, V. P., Papisov, I. M.: Vysokomol. Soedin. **3,** 426 (1961)
147) Natta, G., Mazzanti, G., Pregaglia, G. F., Pozzi, G.: J. Polymer Sci. **58,** 1201 (1962)
148) Gandini, A., Rieumont, J.: Brit. Polymer J., in press
149) Kunitake, T., Yamaguchi, K., Aso, C.: Makromol. Chem. **172,** 85 (1973)
150) Durgaryan, A. A., Terlemezyan, Z. N., Kirakosyan, Z. A., Sarkisyan, G. S.: Vysokomol. Soedin. A **10,** 303 (1968)
151) Kulakov, V. V., Kamenskii, I. V.: Tr. Mosk, Khim. Techn. Inst. **66,** 166 (1970); C.A. **75,** 77367 (1971)
152) Girdyuk, V. V., Kirilenko, Y. K., Volf, L. A., Meos, A. I.: Zh. Prikl. Khim. **39,** 2601 (1966)
153) Kamenskii, I. V., Filimonova, S. M., Van Lam, N., Yerishev, B. Y., Kroviakova, N. B.: Plast. Massy **1973** (5), 5
154) Girdyuk, V. V., Kirilenko, Y. K., Volf, L. A., Meos, A. I.: Zh. Prikl. Khim. **40,** 1386 (1967)
155) Shelkunov, N. G., Klimenko, I. B., Girdyuk, V. V., Volf, L. A.: Khim. Geterots. Soedin. **5,** 775 (1969)
156) Tsuda, M.: J. Polymer Sci. Al **7,** 259 (1969)
157) Borrows, E. T., Haward, R. N., Porges, J., Street, J.: J. Appl. Chem. **5,** 379 (1955)
158) Hernandez, C., Galego, N., Llerandi, N., Gandini, A.: Revista CENIC, Sci. Fis. **5,** 217 (1974); C.A. **82,** 98545 (1975)
159) Clarke, J. T., Howard, R. O., Stockmayer, W. H.: Makromol. Chem. **44/46,** 427 (1961)
160) Savranskaya, S. D., Trubitsyna, S. N., Askarov, M. A.: Vysokomol. Soedin., Karbots. Vysokomol. Soedin., Sb. Statei, 1963, p. 20
161) Gandini, A., Rieumont, J.: unpublished results
162) Kispert, L. D., Quijano, R. C., Pittman, C. U., Jr.: J. Org. Chem. **36,** 3837 (1971)
163) Pennanen, S., Nyman, G.: Acta Chem. Scand. **26,** 1018 (1972)
164) Pennanen, S.: Acta Chem. Scand. **26,** 1961 (1972)
165) Alvarez, R., Gandini, A., Martinez, R., Ortiz, P. J., Pérez, C. S.: Revista CENIC, Sci. Fis. **5,** 2 (1974); C.A. **82,** 111867 (1975)
166) Pennanen, S.: Acta Chem. Scand. **26,** 1280 (1972)
167) Nakamura, Y., Saito, M.: Kogyo Kagaku Zasshi **62,** 1178 (1959); C.A. **57,** 8523 (1962)
168) Gandini, A., Galego, N.: Rev. Latinoam. Quim., in press
169) Nakamura, Y., Saito, M.: Kogyo Kagaku Zasshi **62,** 1168 (1959); C.A. **57,** 13985 (1962)

Received December 13, 1976
M. Gordon (editor)

Chemical Modifications of Fibre Forming Polymers and Copolymers of Acrylonitrile

Genrikh A. Gabrielyan and Zakhar A. Rogovin

Moscow Textile Institute, Department of Chemistry, Fibre Technology, Donskaya 26, Moscow B – 419, USSR

Table of Contents

1. Introduction 98
2. Synthesis and Modification of Fibre-Forming Acrylonitrile Copolymers Containing Reactive Groups 99
   2.1. Copolymers of AN with Diketene (2-Methylene-4-oxooxetane) 99
   2.2. Copolymers of AN with Unsaturated Aldehydes 102
   2.3. Copolymers of AN with 2-Amino-4-vinylsulfonylanisole 106
   2.4. Copolymers of AN with Diens 107
   2.5. Copolymers of AN with 3-Chloro-2-butenyl Methacrylate 111
   2.6. Copolymers of AN with Vinylpyridines 113
3. Synthesis and Modification of New Acrylonitrile Polymers and Copolymers with the Use of Polymer-Analogous Transformations of the Nitrile Group 115
   3.1. Transformation of the Nitrile Groups of PAN into Aldehyde Groups and Derivatives 115
   3.2. Transformation of the Nitrile Groups of PAN into Amino Groups; Thioamidation 116
4. Synthesis of Block and Graft Copolymers of Polyacrylonitrile 126
   4.1. Graft Copolymers of PAN *via* Diazonium Salts 126
   4.2. Graft Copolymers of PAN with the Use of Redox Systems 127
   4.3. Graft Copolymers of PAN with the Use of Oligomeric Diisocyanates . . 129
   4.4. Block Copolymers of PAN 130
5. References 133

1. Introduction

The great interest in the methods of modification of polymers which has increased during the past decades does not result only from the practical significance of this problem, but also from the fact that, due to the broad application of these methods, potentialities of the modern chemistry of polymers have expanded considerably and it has become possible to create polymers with required properties.

Among the various types of polymers the problem of chemical modification of polyacrylonitrile[a] (PAN) and polyacrylonitrile fibres attracts the ever growing attention of numerous investigators.

The interest in this problem arises from the desire to eliminate some drawbacks typical of PAN and the materials obtained from it: enhanced stiffness, poor dyeability, electrifiability, combustibility, etc., as well as from a wide range of possibilities in producing, based on PAN, new readily available materials with specific technologically valuable properties.

In the overall problem of PAN modification the following methods of chemical modification have acquired great importance:
1. The synthesis of copolymers,
2. the conversion of functional groups in the polymer macromolecules (polymer-analogous reactions),
3. the synthesis of block and graft polymers.

In our laboratory this kind of research has been systematically carried out for a number of years. A certain progress has been made in developing new promising methods for the chemical modification of PAN and acrylonitrile (AN) copolymers, and in creating new types of fibres with technologically valuable properties.

Of considerable scientific and certain technological interest are the methods of synthesizing AN copolymers with monomers containing reactive groups. The use of diketene (DK) (2-methylene-4-oxooxetane)[1], methacrolein (MA)[2], 2 amino-4-(vinylsulfonyl) anisole (VSA)[3], or 3-chloro-2-butenyl acrylates (CCA)[4] as the second monomer has made it possible to synthesize fibre-forming AN copolymers containing lactone, aldehyde, amino, or 3-chloro-2-butenyl groups. Investigations, recently carried out, have shown the possibility of using dienes[5] and quaternary salts of 2-methyl-5-vinylpyridine[6] for the synthesis of fibre-forming AN copolymers.

The insertion of the above types of functional groups into the macromolecules of AN copolymers is of substantial interest for carrying out a variety of subsequent conversions:
1. For the insertion of new types of functional groups which cannot be directly obtained by synthesizing copolymers from monomers,
2. for effecting crosslinkage,
3. for the synthesis of graft copolymers, etc.

Using classical reactions of nitriles of low molecular weight[7] a number of conversions can be effected in the PAN polymer chain ontaining new derivatives. The reduction of the nitrile groups and subsequent Stephen reaction followed by

---

[a] Systematic name: poly(1-cyanoethylene).

thioamidation is of greatest interest from the point of view of studying the distinguishing features of the polymer analogous reactions of PAN, as well as obtaining fibres with special properties.

This paper presents the main scientific results obtained by using these methods of chemical modification of PAN.

2. Synthesis and Modification of Fibre-Forming Acrylonitrile Copolymers Containing Reactive Groups

Copolymerization is a classical method of PAN modification which has acquired the greatest scientific and technological importance.

As it is known at present in all countries producing PAN fibres, AN copolymers containing 5–10% of a second monomer are used to increase the elasticity and, in most cases, a third monomer (1–25%) is added to improve the dyeability.

2.1. Copolymers of AN with Diketene (2-Methylene-4-oxooxetane)

More and more emphasis has recently been placed on the synthesis of fibre-forming AN copolymers containing reactive groups[8]. Among the monomers used to obtain reactive copolymers of considerable interest is diketene (1) (2-methylene-4-oxooxetane).

$$CH_2=C\underset{O}{\overset{CH_2}{<}}C=O$$

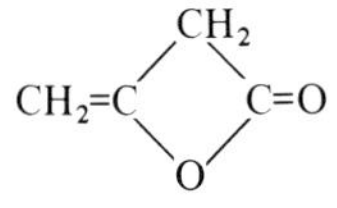

1

As early as 1940 it has been established[9] that diketene does not polymerize by a radical mechanism. It has, however, been shown later[10] that it undergoes reactions of radical copolymerization with many vinyl monomers[11]. In this reaction the double bond is involved and the lactone ring is preserved in the copolymer.

Thus, the copolymerization of AN with *1* yields polymers of type *2*[b], containing β-lactone rings[12].

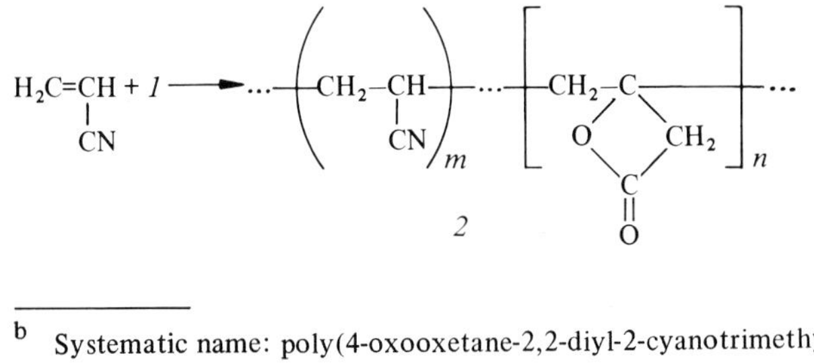

$$H_2C=CH + 1 \longrightarrow \cdots -(-CH_2-CH-)_m-(-CH_2-C\underset{\underset{O}{\overset{\|}{C}}}{\overset{O}{<}}CH_2-)_n- \cdots$$
$$\quad\quad|\quad\quad\quad\quad\quad\quad\quad\quad|$$
$$\quad CN\quad\quad\quad\quad\quad\quad CN$$

2

---

[b] Systematic name: poly(4-oxooxetane-2,2-diyl-2-cyanotrimethylene).

Although there are indications[13] that copolymers of AN with *1* containing 5–10% of *1* are obtained in an inert solvent at 80 °C with a monomer ratio of 1 : 1, and even in aqueous solution[14] in which *1* is known[15] to dissolve readily forming acetone and CO_2, fibre forming copolymers can only be obtained under comparatively mild conditions[1] in benzene solution. *1* is a much less reactive monomer than AN. The reactivity ratios found in the copolymerization in benzene are $r_1 = 4{,}23 \pm 0{,}06$ for AN and $r_2 = 0$ for *1*. A number of authors [15] indicate that in the copolymerization in mass these values amount to $r_1 = 5{,}89$ and $r_2 = 0{,}06$ respectively. There are also indications that $r_1 = 7{,}0$ and $r_2 = 0$[16].

Since *1* is a monomer with low activity, copolymers *2* obtained at any stage of the copolymerization process, irrespective of the monomer ratio in the initial mixture, always contain a smaller amount of monomeric units of *1* than that in the corresponding monomer mixture. *1* being prone to enter the chain-transfer reaction, the increase of its content in the initial monomer mixture reduces substantially the reaction rate and decreases the molecular mass of the copolymers. It was found that copolymers *2* which contain 2–8% of monomeric units of *1* and are suitable for obtaining fibres must have a molecular mass between 45 000 and 50 000. Such copolymers can be obtained with a AN: *1* ratio in the initial mixture between 95 : 5 and 85 : 15. Concentrated solutions of copolymers, especially those with a molecular mass smaller than the above limit, are characterized by a very low stability which is a substantial shortcoming of these copolymers.

Fibres spun out of copolymers *2* by the wet process from dimethylformamide (DMF) solutions have the following characteristics: strength 23–27 g.f./tex (2,5–3 g/denie), elongation 15%.

The presence of β-propiolactone rings in the macromolecules of copolymers *2* makes it possible to accomplish, both in homogeneous medium in DMF and in the finished fibre, numerous reactions known for these type of compound. Depending on the reaction conditions, in particular on the character of the solvent and the chemical reagent, the opening of the lactonic ring occurs as in the case of low-molecular lactones, either at the bond between the carbonylcarbon and the oxygen or between the alkylcarbon and the oxygen. Reactions of the interaction of copolymers containing 6–10% of monomeric units of *1* with amines or with thioglycolic acid have been studied, as well as the hydrolysis of lactone rings in the presence of sulfuric and phosphoric acid.

The study of the functional composition of the newly obtained PAN derivatives has shown that during the interaction with ammonia or amines (hydroxylamine, 2,2'-iminodiethanol, etc.) the opening of the lactone ring is taking place at the carbonylcarbonoxygen bond, which results in the formation of units containing hydroxyl and amido or substituded amido groups. Solutions of thioglycolic acid in acetone react with these copolymers with the splitting of the alkylcarbon-oxygen bond. In this case only carboxylic groups are introduced into the modified PAN fibre. These polymer-analogous conversions of copolymers *2* are shown in Scheme 1. (See page 101.)

As a result of the interaction with bifunctional compounds, *e.g.*, with hexamethylenediamine, disubstituted amino groups are formed providing chemical bonds

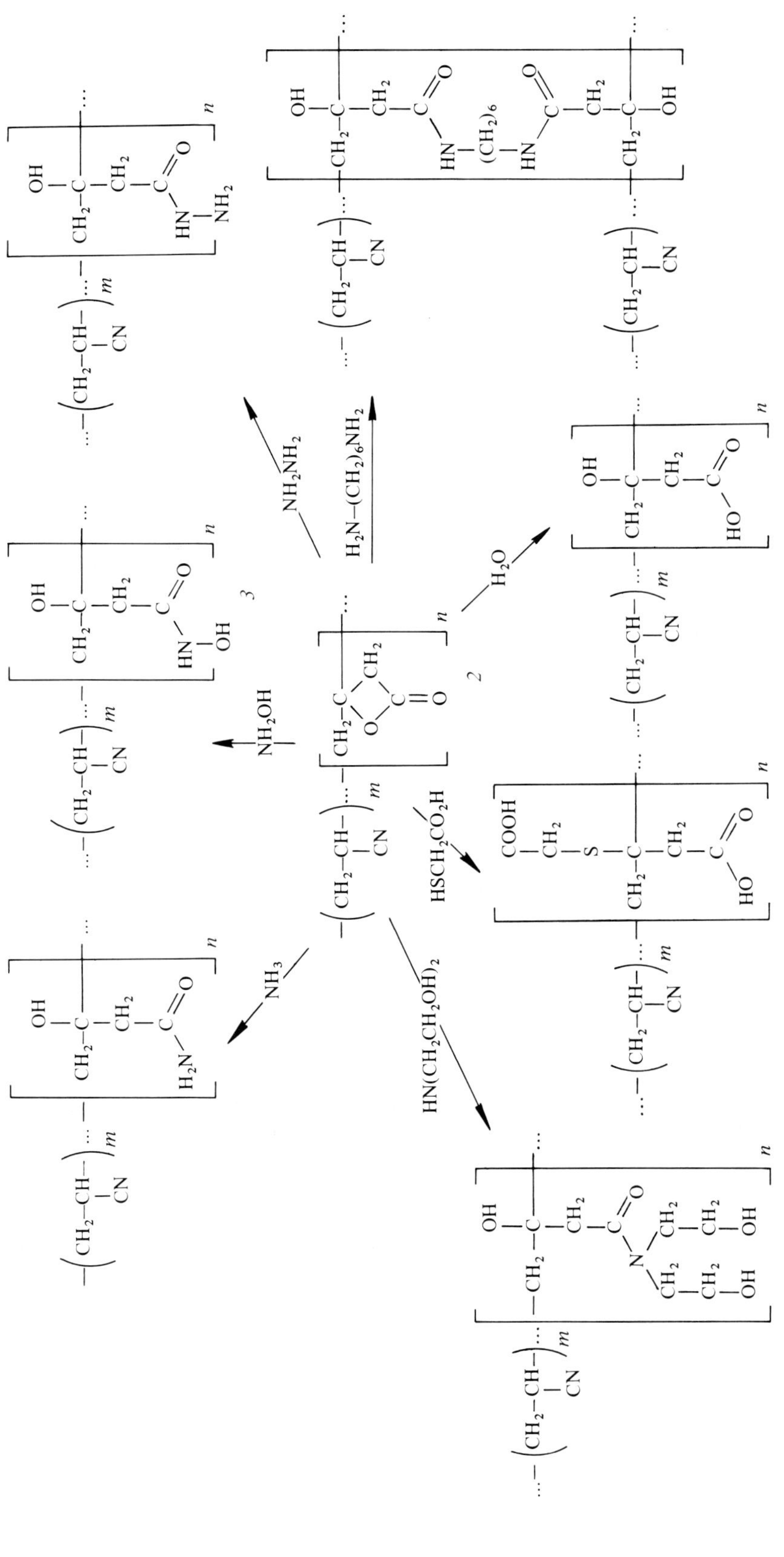

Chemical Modifications of Fibre Forming Polymers and Copolymers of Acrylonitrile

Scheme 1

between the macromolecules, which results in a considerable increase of the heat resistance of the fibres[17]. Modified fibres are stained with acid and active dyes.

It should be noted that with dyes containing amine groups, e.g., dispersed 1-aminoanthraquinone ("red GS"), chemically stained fibres are obtained[17]:

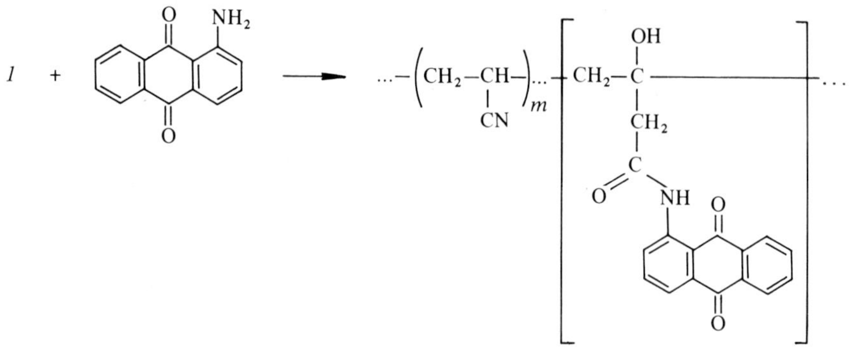

By transformation of modified PAN fibres of type *2* into *N*-hydroxyamide derivatives of type *3* (Scheme 1), it is possible to form intermolecular chelate bonds by interaction with Fe^{+3}:

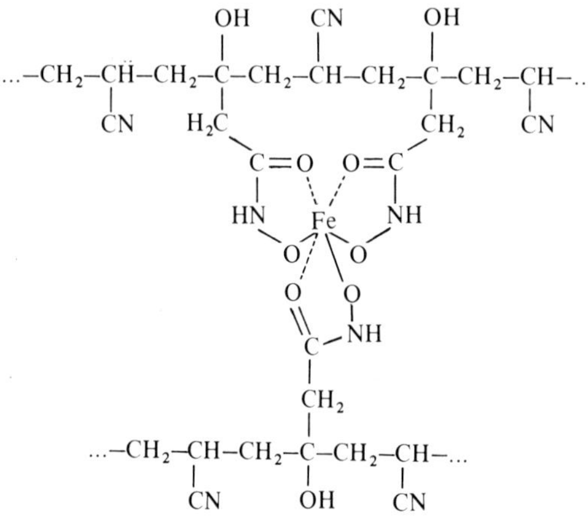

The presence of these bonds contributes to the increase of the temperature of zero strength from 140 °C for PAN to 280–320 °C, as well as to a lower shrinkage at high temperatures.

2.2. Copolymers of AN with Unsaturated Aldehydes

Of substantial interest for the synthesis of fibre-forming reactive AN copolymers is the use of methacrolein *4* as second monomer, which affords copolymers of type *5*, containing aldehyde groups.

$$H_2C=CH + H_2C=\overset{CH_3}{\underset{|}{C}}-C\overset{\nearrow O}{\underset{\searrow H}{}} \longrightarrow \cdots\left(CH_2-\underset{\underset{CN}{|}}{CH}\right)_m \cdots\left(CH_2-\underset{\underset{CHO}{|}}{\overset{\overset{CH_3}{|}}{C}}\right)_n\cdots$$
$$\text{CN}$$
$$\hspace{3cm} 4 \hspace{6cm} 5$$

The insertion of these groups offers a variety of possibilities[18] for further modifications of the fibres.

The process of copolymerization of AN with 4 has been studied in aqueous solution (in suspension), in emulsion, and in dimethylformamide (DMF) solution[3,19].

The reactivity ratios for this pair of monomers, found[19] for the copolymerization in mass, are 0,06 for AN and 2,0 for 4.

The values of these ratios change appreciably by passing from the heterogeneous (suspension) to the homogeneous (DMF) system. In the case of copolymerization in suspension in the presence of the $K_2S_2O_8-AgNO_3$ oxidation-reduction system at 30–40 °C, the ratios were found to be $r_1 = 0,77 \pm 0,2$ and $r_2 = 1,09 \pm 0,04$, whereas in the case of the copolymerization in solution they are $r_1 = 0,52$ and $r_2 = 1,7$. The difference in these values seems to be the result of the different solubility of the monomers in water and of the different rate of diffusion of the monomers to the surface of the precipitated copolymer[20]. From this it follows that 4 is the more reactive monomer in this binary system.

During the copolymerization 4 is, therefore, consumed faster, and at the initial stage the copolymers are enriched with 4. At high conversion levels its relative portion in the macromolecules of the copolyner decreases, resulting in the formation of copolymers with non-uniform composition.

Addition of 4 to AN was found to decrease considerably both the polymerization rate and the molecular mass of the polymers. This effect is, possibly, caused by a comparatively low reactivity of the radicals formed from AN and, on the other hand, by the aldehyde groups being prone to form radicals and their participation in the transfer reactions of the growing chain. This makes the system especially sensitive to temperature. The reaction rate increases with increasing copolymerization temperature, however, together with the rate of chain interruption and chain transfer through the monomer[3] and, correspondingly, it decreases considerably the specific viscosity of the copolymer solutions.

The number of aldehyde groups in the copolymers of type 5 has been found to be always smaller than that of cyanogroups. This is a characteristic feature of the copolymerization of AN with 4. An especially great difference is observed when the copolymerization reaction is carried out in benzene and DMF at elevated temperatures. In this case the content of aldehyde groups amounts to only 50–60% of the amount of aldehyde groups calculated from the nitrogen content in the copolymers[19].

It has been shown by Schulz and Kern[21] that the radical polymerization of acrolein can take the course of the 1,2-mechanism as well as that of the 1,4- or 3,4-mechanism leading to formylethylene, oxy-2-propenylene, or oxy-2-propenylidene units, respectively. This behaviour of acrolein and its derivatives seems to be also retained to a certain extent, in the radical copolymerization of 4 with AN causing a decrease fo the content of aldehyde groups in the copolymers.

It is, however, possible that the difference, to a certain extent, can be explained by the participation of the aldehyde groups in the chain-transfer reaction:

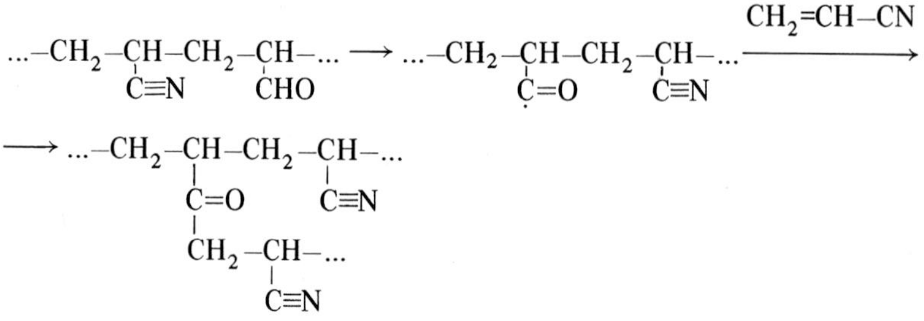

$$\ldots-CH_2-CH-CH_2-CH-\ldots \longrightarrow \ldots-CH_2-CH-CH_2-CH-\ldots \xrightarrow{CH_2=CH-CN}$$
$$\underset{C\equiv N}{|}\underset{CHO}{|} \underset{C=O}{|}\underset{C\equiv N}{|}$$

$$\longrightarrow \ldots-CH_2-CH-CH_2-CH-\ldots$$
$$\underset{C=O}{|}\underset{C\equiv N}{|}$$
$$\underset{CH_2-CH-\ldots}{|}$$
$$\underset{C\equiv N}{|}$$

Copolymers of homogeneous composition containing 5–8% of monomeric units of *4*, 0,5% solutions in DMF having a specific viscosity of 1,3–1,5, and being suitable for obtaining fibres, can be synthesized using a AN/*4* mole ratio of 99:1 by suspension copolymerization with stage addition of monomers in the presence of a small amount (0,8–1%) of 2-aminoethanol as chain growth regulator[3].

The fibres spun from copolymers of AN with *4* formed by the wet process from DMF solutions have satisfactory physico-mechanical properties.

Modified PAN fibres containing aldehyde groups can be used to obtain chemically stained fibres. The chemical addition of dyes can be conducted following two schemes:

A) The fibres of the above composition are treated with aromatic amines, e.g., 1-amino-8-hydroxynaphthalene-3,6-disulfonic acid ("H-acid"), and then azocoupling is carried out with a diazonium salt:

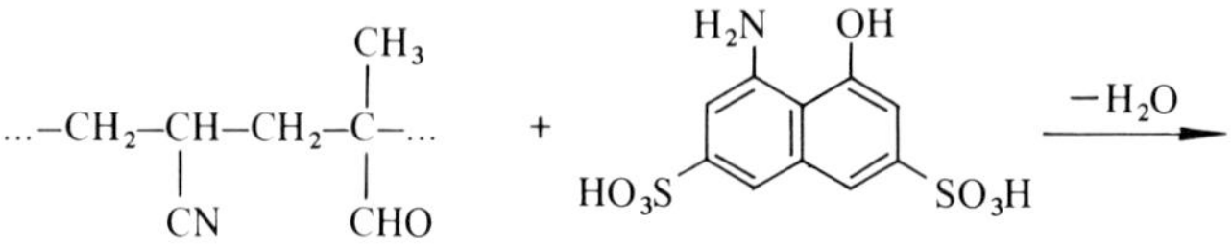

5

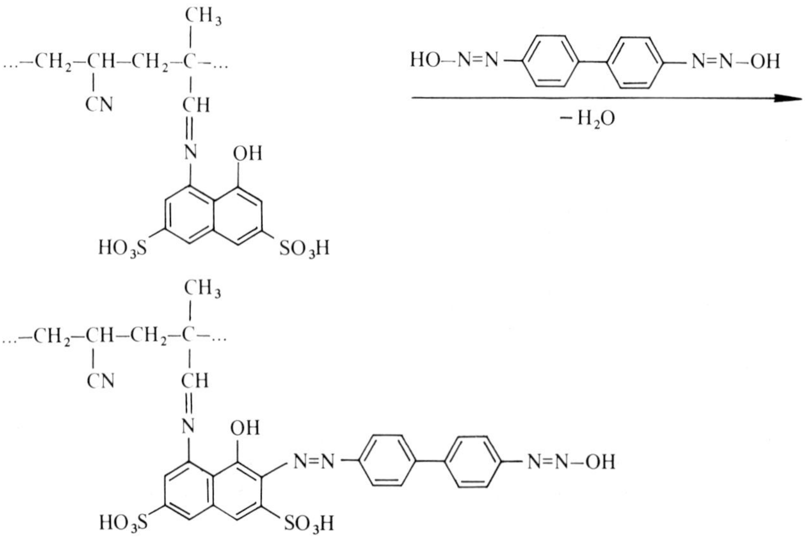

B) The aldehyde groups interact with dyes containing amino groups:

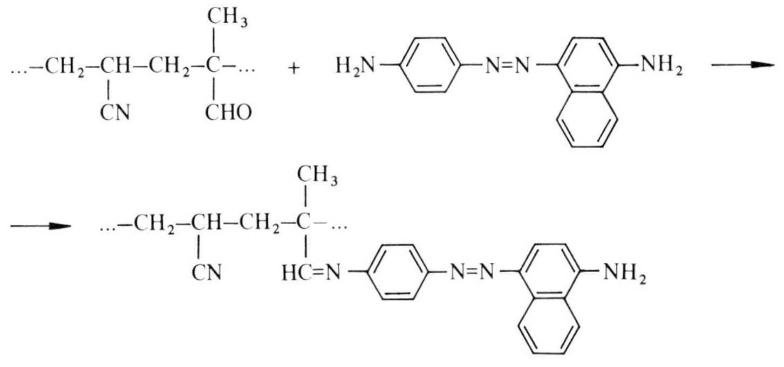

$$...-CH_2-CH(CN)-CH_2-C(CH_3)(CHO)-... + H_2N-C_6H_4-N=N-C_{10}H_5-NH_2 \longrightarrow$$

$$\longrightarrow ...-CH_2-CH(CN)-CH_2-C(CH_3)(HC=N-C_6H_4-N=N-C_{10}H_5-NH_2)-...$$

The degree of substitution of aldehyde groups does not exceed 25%, the structure of the dye having no effect on the degree of substitution.

When modified fibres of type 5 are treated with hydroxylamine, oxime groups are also easily formed. The interaction with a protein affords a sandwich polymer[22]. Fibres modified in this way have enhances dyeability. When copolymer fibres are treated with diamine solutions or in acid medium with Fe^{+3} salts, intermolecular chemical bonds are formed, which results in a considerable increase of the temperature of zero strength and of the heat resistance of fibres. These conversions are shown in Scheme 2.

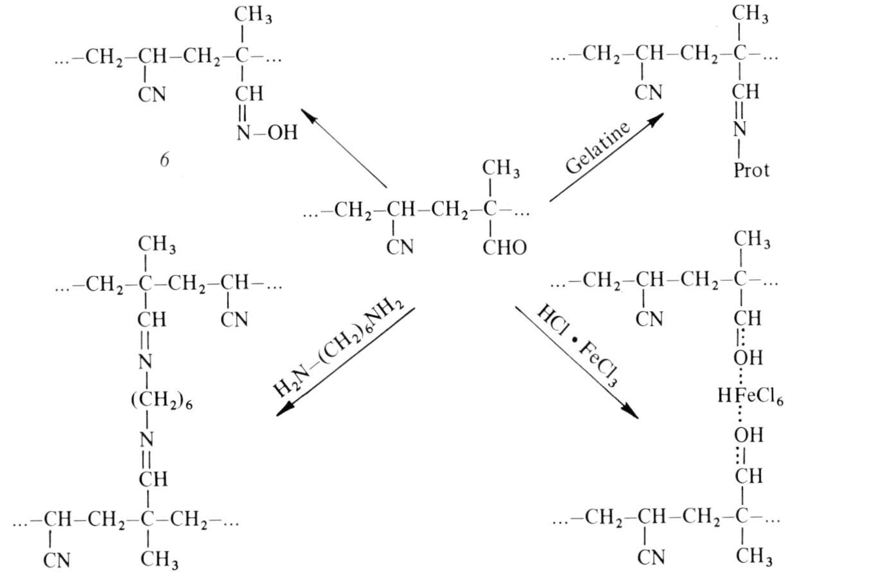

Prot.: protein residue
Scheme 2.

It should be noted that fibres with high heat resistance can be also obtained by treating modified fibres of type 6, containing hydroxyimino groups, with Fe^{3+} and Ni^{2+} salts, which is again explained by the formation of intermolecular chemical bonds.

2.3. Copolymers of AN with 2-Amino-4-vinylsulfonylanisole

Of substantial interest for obtaining reactive fibres capable of being chemically stained as well as subsequently subjected to the reaction of graft copolymerization are the copolymers of AN with 2-amino-4-vinylsulfonylanisole 7.

7 does not polymerize by the radical mechanism. It, however, enters the radical copolymerization with AN which results in the formation of a copolymer with structure 8[23]:

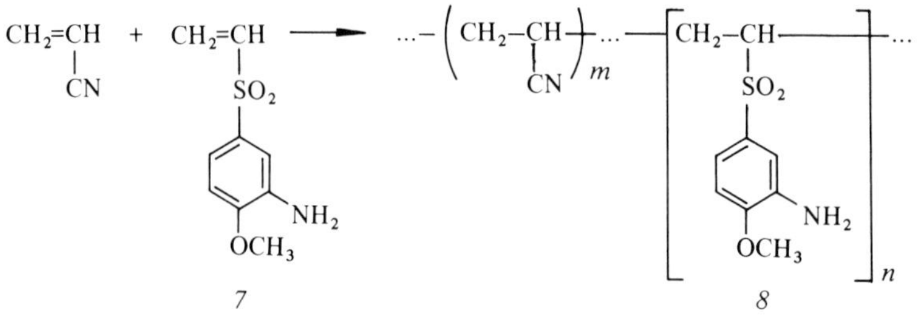

$$CH_2=CH(CN) + CH_2=CH(SO_2-C_6H_3(NH_2)(OCH_3)) \longrightarrow \cdots-(CH_2-CH(CN))_m-[CH_2-CH(SO_2-C_6H_3(NH_2)(OCH_3))]_n\cdots$$

7　　　　　　　　　　　　　　　　8

It has been found[24] that copolymerization does not proceed in the presence of benzoyl peroxide, which seems to be due to the interaction of benzoyl peroxide with aromatic amines[25]. 2,2'-azoisobutyronitrile was used as initiator.

Reactivity ratios for the copolymerization of AN with 7 in methanol at 60 °C, proved to be equal to $r_{1_{AN}}= 3{,}6 \pm 0{,}2$ and $r_{2_{7}} = 0 \pm 0{,}06$, i.e., AN is a much more active component in this binary system. The low reactivity of the vinyl double bond in 7 is explained by the specific effect of the sulfonyl group on its polarity[26]. In addition to that, the radical formed from 7 does not seem to be stabilized by the sulfonyl group and readily takes part in the chain transfer reaction and chain termination. As a result of this, the rate of copolymerization reaction and the molecular mass of the copolymers decrease with increasing content of 7 in the initial mixture.

Investigations on the copolymerization kinetics for the ratio of AN : 7 = 97,5 : 2,5 showed that the process adheres to the general laws governing the polymerization and copolymerization of AN in suspension[27].

It was established that for the further modification of copolymers, as well as that of finished fibres, using the aromatic amino group, and, in particular, to achieve deep staining, a content up to 2% of monomeric units of 7 in the copolymer is sufficient.

The optimum molecular mass (M_w) of these copolymers containing 1,5–2% of monomeric units of 7, amounts to 47 000–55 000 which makes it possible to obtain concentrated (15–17%) solutions with sufficiently high stability and to spin fibres with high physicomechanical properties (P = 36–40 gf/tex, (4–4,4 g/denie), l = 12–15%).

Chemical staining of fibres of type 8 containing units of 7 is performed by diazotation followed by azocoupling with "H-acid" or with β-naphthol as shown by the following scheme:

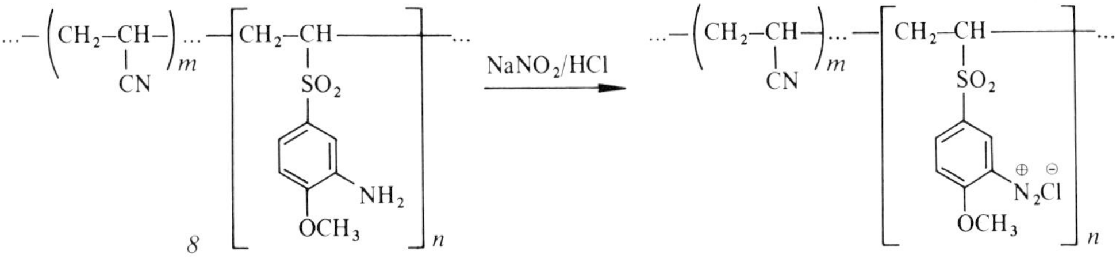

Chemical Modifications of Fibre Forming Polymers and Copolymers of Acrylonitrile

[Reaction scheme: PAN-SO₂-aryl(NH₂)(OCH₃) copolymer treated with NaNO₂/HCl to give the diazonium salt N₂Cl derivative; compound **8**. Then coupling with HO,NH₂-naphthalene-SO₃H/HO₃S reagent gives the azo-coupled copolymer with -N=N- linkage.]

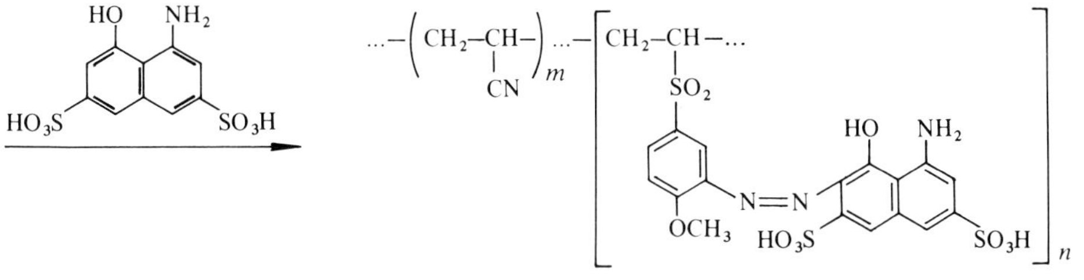

As it will be shown the presence of an aromatic amino group in the macromolecule of modified PAN offers new possibilities for graft copolymers without the formation of a homopolymer.

2.4. Copolymers of AN with Diens

Copolymers of AN with diene monomers and, in particular, with butadiene and isoprene, deserve special attention. These copolymers with a predominating content of monomeric units of dienes are known to have been produced in the form of rubbers for a long time and are finding a broad application in various branches of technology.

However, no studies have been carried out until recently on the synthesis of AN copolymers containing only a small quantity of monomeric diene units which may have fibre forming properties. It is only in the last few years that several reports have appeared on the copolymerization of AN with butadiene in DMF[28] and on the use of AN-butadiene copolymers to obtain fibres[29].

The synthesis of copolymers from AN and isoprene (ISP) is one of the most promising and effective methods of modifying the properties of PAN to obtain fibres with improved practical properties[40]. This fibre-forming AN copolymer can, in principle, be also used to spin fibres without using a solvent.

As it has been shown lately, insertion of a small quantity (5–15% of the copolymer weight) of ISP monomeric units into the PAN macromolecules results in an appreciable decrease of stiffness and in an increase of flexibility of the chain, which makes it possible to improve considerably the fatigue properties of usual PAN fibres[30]. In addition to that, by inserting a comparatively large amount (25–30%) of flexible ISP monomeric units into the copolymer one can decrease substantially the yield temperature of PAN, which makes it possible to spin fibres from thermoplastic state[31].

The presence of double bonds in the macromolecule of these copolymers of type 9 makes it possible to crosslink the fibres by various methods[32, 33].

$$\cdots\left(-CH_2-CH\atop CN\right)_m\cdots\left(-CH_2-\underset{CH_3}{C}=CH-CH_2-\right)_n$$

9

Fibre forming copolymers of type 9 have been synthesized in dimethyl sulfoxide (DMSO) and in emulsion[34].

During the investigation of the principles governing the process of copolymerization of AN with ISP in DMSO at 30 °C in the presence of ammonium persulfate, it was established that the anisotropic type of copolymerization is characteristic for this pair of monomers. The azeotropic point, as it is seen from Fig. 1 corresponds to a content of 60% of monomeric units of ISP in the monomer mixture.

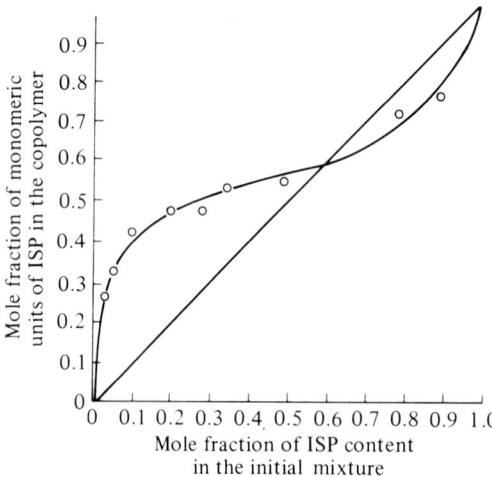

Fig. 1. Dependence of the composition of the copolymer from AN and ISP on the monomers ratio in the initial mixture

The reactivity ratios for the copolymerization of AN with ISP, determined under the above reaction conditions, were equal to $r_{1\,AN} = 0{,}052 \pm 0{,}004$ and $r_{2\,ISP} = 0{,}38 \pm 0{,}08$.

The values of r_1 and r_2 found in the block copolymerization for this pair are equal to $r_1 = 0{,}03 \pm 0{,}03$ and $r_2 = 0{,}45 \pm 0{,}05$ respectively[35]. The difference in these values indicates that DMSO produces a noticeable effect on the reactivity of AN.

There is some reason[36] to believe that this noticeable change in the reactivity of AN in DMSO solution, is caused by the dipole-dipole interaction of the nitrile group of AN with the sulfinyl group of the DMSO molecule, resulting in an appreciable change in the polarity of the double bond of AN as well as in the activity of the acrylonitrile radical. It is one of the characteristic features of the AN-ISP binary system that the addition of even small quantities (1–2%) of ISP to AN leads to a sharp decrease of the polymerization rate of AN (Fig. 2).

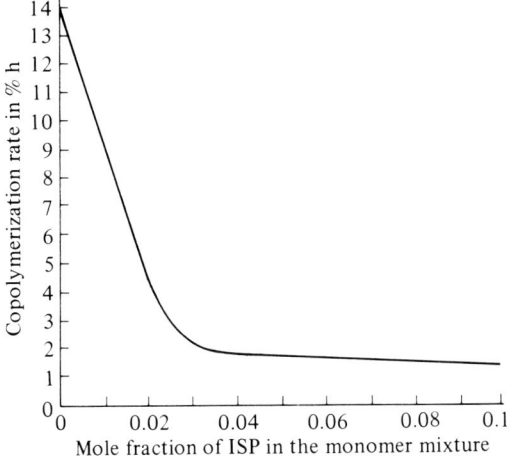

Fig. 2. Dependence of initial copolymerization rate on the ISP content in the initial monomer mixture

This is explained by the fact that, as a result of ISP addition to the growing chain containing at its end a cyanomethylene radical, an allyl radical is formed at the end of the chain:

$$\ldots-CH_2-\overset{\cdot}{\underset{CN}{CH}} + CH_2=\overset{CH_3}{\underset{|}{C}}-CH=CH_2 \longrightarrow \ldots-CH_2-\underset{CN}{\overset{|}{CH}}-CH_2-\overset{CH_3}{\underset{|}{C}}H=CH-\overset{\cdot}{C}H_2$$

The allyl radical is better stabilized by resonance with the adjacent double bond than the cyanomethylene radical and is, therefore, less reactive.

The kinetic order of the copolymerization reaction with respect to the initiator is equal to 0,5, and the total activation energy amounts to 14,4 kcal/mol (60,3 kJ/mol).

It is a characteristic feature of this copolymerization, as in general of binary systems in which one of the monomers is a diene, that with the progress of the reaction a secondary crosslinking process becomes possible.

Since the double bonds in the macromolecules of the copolymer have a much lower activity than that of the vinyl groups in the ISP molecule, crosslinkage occurs in this case at later stages of reaction, i.e., when the monomer conversion exceeds 50%.

The insertion of ISP monomeric units also results in significant changes in the thermomechanical properties of the copolymers. The glass-transition temperature of PAN is equal to 95 °C, whereas if 15–21,5% of ISP were added to the copolymer, this temperature decreases to 78 and 63 °C, respectively. Besides this, the exothermal effect, characteristic of PAN and caused by the cyclization reaction taking place, is also reduced considerably[36].

Modified PAN fibres have been obtained from copolymers containing up to 15% or ISP units using the wet spinning process[30]. Some properties of modified fibres are presented in Table 1. For comparison are also given the properties of fibres obtained from copolymers additionally crosslinked with conventional crosslinking agents used in the vulcanization of nitrile rubbers.

Table 1. Physicomechanical properties of modified PAN fibres[a]

Fibre	Crosslinking agent[b]			Strength in gf/tex[c]	Elongation in %	Abrasive resistance, no. of cycles	Resistance to double bend, no. of cycles
	thiuram D	thiuram + sulfur	diphenyl guanidine + sulfur				
PAN	–	–	–	35,8	16,2	114	122
PAN	+	–	–	31,4	14,3	208	120
	–	–	–	32,4	16,1	150	395
ISOPAN-4	+	–	–	32,2	25,2	280	1065
	–	+	–	32,5	18,8	390	1532
	–	–	–	26,6	19,6	240	630
ISOPAN-8	+	–	–	26,4	32,3	893	1715
	–	+	–	26,2	36,6	1280	1995
	–	–	+	26,4	37,3	1090	1880
ISOPAN-15	–	–	–	27,4	20,0	524	710
ISOPAN-15	+	–	–	27,0	40,0	1300	2100

[a] Fibres from copolymers are denoted as "ISOPAN", figures indicate ISP content in the copolymer in %.
[b] The sign + means that the crosslinking agent was used, i.e. introduced into the spinning solution. Crosslinkage was conducted at 140 °C. Thiuram D: tetramethylthiuramdisulfide.
[c] $gf/tex := \frac{1}{9} g/denie$.

As it can be seen from the above data, by introducing 4–15% of monomeric units of ISP into the macromolecules of the AN copolymer, the elastic properties of PAN fibres and, especially, their resistance to abrasion and double bends can considerably be improved.

Copolymers of AN with ISP, containing more than 25% of monomeric units of ISP with a molecular mass of 50000 to 60000, obtained in emulsion at pH 3, in distinction to PAN, are capable of passing into the state of viscous flow without destruction and cyclization and are processed into fibres at 180–220 °C. When copolymers of higher molecular mass are used it is necessary to raise the temperature of processing. This leads to an intensive crosslinking and to cyclization, due to which it becomes impossible to obtain fibres from them.

Fibres spun from copolymers in the thermoplastic state and subjected to thermal stretching by 500–1000% have the strength of 15–25 g.f./tex. However, these fibres, as all the fibres spun in softened state from AN copolymers containing 25–30% of the second monomer forming a flexible-chain polymer, e.g., isobutylene[37], acrylates[38], vinyl acetate[39], etc., are characterized by a low stability of form and a high shrinkage (60–70%) in boiling water.

This serious disadvantage of the above types of modified PAN fibres, which excludes the possibility of their practical utilization, can be removed by subsequent crosslinkage.

The presence of double bonds in the macromolecule of copolymers creates the necessary conditions for the subsequent implementation of crosslinkage of the fibres obtained.

The results obtained make it possible to conclude that, to solve successfully the problem of spinning modified PAN fibres from softened state without using solvents, of all the great number of copolymer proposed for this purpose, fibre-forming AN copolymers with diene monomers are the most promising ones.

2.5. Copolymers of AN with 3-Chloro-2-butenyl Methacrylate

Much attention has recently been paid to the questions of obtaining PAN fibres with enhanced heat resistance. Unfortunately, in most cases the chosen crosslinking agents as well as AN copolymers containing functional groups capable of crosslinking, did not produce the desired results. These crosslinking agents or new functional groups inserted into the macromolecules of PAN to achieve crosslinkage along with the formation of intermolecular chemical bonds, as a rule, also facilitate the reaction of cyclization. This makes it impossible to assess unambiguously the effect of cross-linkage on the properties of these fibres, in particular on their heat resistance and thermal stability. Therefore, the synthesis of new types of fibre-forming AN copolymers with such monomers which undergo selective crosslinkage of fibres, without nitrile groups taking part in the reaction of crosslinking and cyclization, is of considerable practical and theoretical interest.

Copolymers of AN with 3-chloro-2-butenyl methacrylate (10) are of interest in this respect.

In distinction to other esters of acrylic acids containing double bonds in the alcohol radical and, therefore exhibiting a tendency to cyclopolymerization[43] and formation of crosslinked polymers, 10 reacts with AN in DMF solution[41] or in benzene/DMF[42] only with the vinyl group of the acid part due to deactivation of the double bond in the 3-chloro-2-butenyl group by the chlorine atom. The copolymer of structure 11 is formed.

$$CH_2=CH + CH_2=\underset{\underset{\underset{O-CH_2-CH=C-CH_3}{|}}{\underset{Cl}{|}}}{\underset{C=O}{\overset{CH_3}{\overset{|}{C}}}} \longrightarrow \ldots-CH_2-CH-\ldots-CH_2-\underset{\underset{\underset{O-CH_2-CH=C-CH_3}{|}}{\underset{Cl}{|}}}{\underset{C=O}{\overset{CH_3}{\overset{|}{C}}}}-\ldots$$

$$\underset{C \equiv N}{} \qquad\qquad\qquad\qquad \underset{C \equiv N}{}$$

10   11

In the copolymerization reaction with AN *10* manifests the same reactivity as that of substituted methacrylates[44]. The value of 0–3 for the polar factor e and the reactivity ratios determined for AN to $r_1 = 0{,}25 \pm 0{,}05$ and for *10* to $r_2 = 1{,}86 \pm 0{,}15$ (in DMFA) and $r_1 = 0{,}18 \pm 0{,}05$, $r_2 = 2{,}12 \pm 0{,}15$ (in benzene) confirm this conclusion. In this binary system, the addition of *10* to AN results in a considerable increase of the polymerization rate.

The fact that crosslinking proceeds at considerably high conversions of the monomers (50%) and at copolymerization temperatures $> 60\,°C$ is a distinctive feature of this system, as compared with systems containing other unsaturated derivatives of acrylic acids, which is determined by the difunctional nature of *10*.

Insertion of monomeric units of *10* into macromolecules of PAN results in the lowering of the glass-transition and softening temperature and also reduces the cyclization intensity of the nitrile groups.

Fibres based on AN copolymers containing 4–10% of monomeric units of *10*[42] and obtained by wet spinning from solutions in DMF have a much better (2–8 times) resistance to multiple deformations than PAN fibres and have a higher light-fastness than PAN fibres. They are, however, inferior to the latter with respect to abrasive resistance and thermal stability.

The presence of the 3-chloro-2-butenyl groups in the macromolecules of these AN copolymers, the structure of these groups being similar to that of the monomeric unit of chloroprene ($-CH_2-CH=\overset{Cl}{C}-CH_2-$), makes it possible to use compounds conventionally applied in the vulcanization of chloroprene rubbers, and in particular zinc oxide, for the crosslinking of these modified PAN fibres.

It has been shown[44] that on heating modified fibres containing ZnO at 150–170 °C, zinc chloride is formed as a result of crosslinkage by interaction of the 3-chloro-2-butenyl groups with ZnO. Under these conditions the nitrile groups of the copolymer molecules do not undergo any chemical conversions.

At present it is believed that intermolecular chemical bonds are formed during the vulcanization of polychloroprene with ZnO not only due to the mobile chlorine in allyl position but also as a result of the reaction of the chlorine located directly at the double bond of the monomeric units chloroprene connected in the chain in 1,4-position as shown in the following scheme[43].

$$\ldots-CH_2-CH=\overset{Cl}{C}-CH_2-\ldots \longrightarrow \underbrace{\ldots-CH_2-CH=\overset{\cdot}{C}-CH_2-\ldots}_{Ka^{\cdot}} + \overset{\cdot}{Cl}$$

Ka is a macromolecule of polychloroprene.

$$Ka\dot{H} + \dot{C}l \longrightarrow Ka\dot{} + HCl \qquad\qquad 2Ka\dot{} \longrightarrow Ka-Ka$$

i.e., crosslinkage is taking place due to thermooxidative dehydrochlorination, and zincs is only used to bind the HCl evolved. Taking into account, however, that no crosslinking of modified fibres containing 3-chloro-2-butenyl groups occurs under the above conditions in the absence of ZnO (the solubility of fibres in DMF is preserved), it can be supposed that it is hardly probable that this reaction should take place during the crosslinkage. It is more probable that crosslinks are formed as a result of the direct interaction with a labile chlorine atom, which is probably formed in the course of an oxidation reaction resulting in the formation of 1-chloroethylcarbonyl groups:

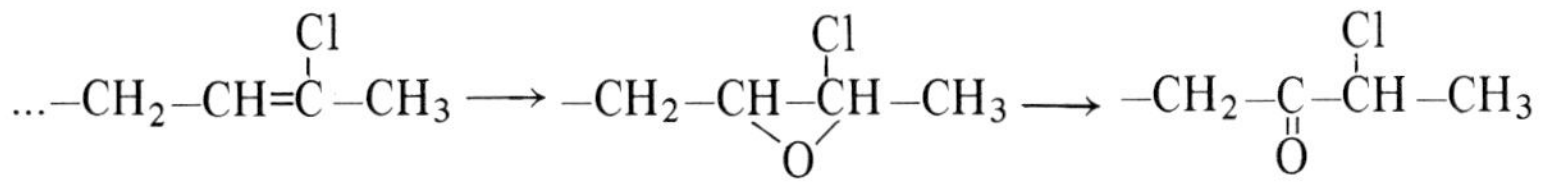

$$...-CH_2-CH=\overset{Cl}{C}-CH_3 \longrightarrow -CH_2-CH\overset{Cl}{\underset{O}{-}}CH-CH_3 \longrightarrow -CH_2-\overset{Cl}{\underset{O}{C}}-\overset{}{CH}-CH_3$$

These groups are capable of reacting with ZnO and forming a chemical bond with the macromolecules of the copolymer as shown in the following scheme.

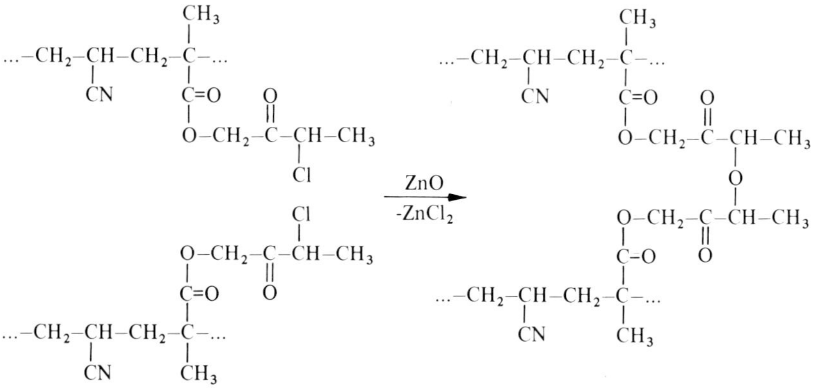

$$\begin{array}{c}
\text{...}-CH_2-CH-CH_2-\underset{\underset{\underset{\underset{\underset{\underset{\underset{CN}{|}}{|}}{|}}{|}}{|}}{C}}{\overset{CH_3}{|}}-\text{...} \\
\end{array}$$

[Reaction scheme showing crosslinking of PAN copolymer with ZnO, eliminating ZnCl₂, to form intermolecular ester/carbonyl crosslinks between two polymer chains]

The results of studying the properties of crosslinked modified PAN fibres showed[44] that crosslinking results in a slight increase of their thermal stability and heat resistance.

These results made it possible to arrive at a sufficiently well-grounded conclusion that the effect of raised heat resistance caused by the formation of intermolecular chemical bonds is not very significant, and that the usually observed considerable increase of heat resistance of PAN fibres as a result of a crosslinkage with bifunctional compounds, is caused not by the formation of intermolecular chemical bonds, as it has usually been thought[45, 46], but by cyclization reactions of the nitrile groups with the formation of naphthyridine cycles[47].

2.6. Copolymers of AN with Vinylpyridines

Of considerable interest are also AN copolymers with vinyl monomers containing amino groups or heterocycles with nitrogen atoms, and, in particular, with vinylpyridines. The synthesis of AN copolymers using monomers of this group is known to have been carried out exclusively with the purpose of imparting stainability to PAN fibres. Recently, however, in connection with the development of research aimed at obtaining ion-exchange fibres, the modification of PAN fibres using such monomers has acquired a new importance.

Of greatest interest for PAN modification with the purpose of obtaining strongly basic anion-exchange polymers can be the quarternary salts of vinyl pyridines.

The distinguishing features of the polymerization and copolymerization to these interesting monomers have been described in a number of papers[48, 49]. There are, however, no systematic investigations described in the literature to study the laws governing the reaction of AN copolymerization with quarternary salts of substituted vinyl pyridines.

In a recently published paper[6], on the investigation of AN copolymerization with the quarternary salt of 1,2-dimethyl-5-vinylpyridinium sulfate (DMVPS) in dimethyl sulfoxide (DMSO) with 2,2'-azoisobutyronitrile as initiator, and in aqueous medium in the presence of the potassium persulfate/sodium metabisulfite oxidation-reduction system at 60 °C, the authors found the reactivity of the monomers, especially that of MVPS (methylvinylpyridin sulfate) to depend significantly on the polarity of the medium.

Copolymerization of AN with the quarternary salt of 1,2-dimethyl-5-vinyl-pyridinium sulfate to proceed in accordance with the following scheme:

$$H_2C=CH \atop | \atop CN \quad + \quad H_2C=CH\text{-pyridinium}(CH_3)\ CH_3SO_4^{\ominus} \quad \longrightarrow \quad ...-CH_2-CH-CH_2-CH-...$$

When passing over from a less polar solvent to a more polar one a polymer enriched with MVPS is formed (Fig. 3).

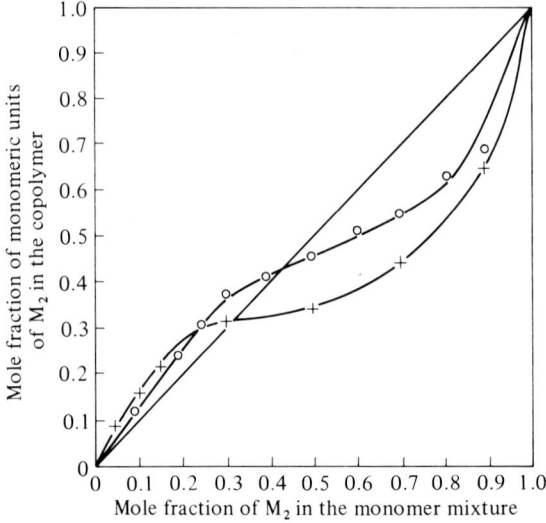

Fig. 3. Dependence of the composition of copolymers of AN with DMVPS (M_2) on the composition of the initial mixture. (o): copolymerization in solution in DMSO; (x): copolymerization in water

Reactivity ratios for the copolymerization of AN and DMVPS in DMSO were found to be $r_1=0.53$ and $r_2=0.036$, and in water $r_1=0.56$ and $r_2=0.25$. The higher reactivity of DMVPS in the copolymerization with AN in aqueous medium, as compared with its reactivity in DMSO, can be explained by a higher degree of dissociation of DMVPS in aqueous medium. This fact also produces a considerable effect on the character of the distribution of monomeric units within the copolymers, which manifests itself in the change of their solubility in water. Copolymers containing 30% of monomeric units AN obtained from a 90:10 mixture of AN and DMVPS in DMSO, irrespective of the level of conversion, are completely soluble in water, whereas copolymers of the same composition, but obtained in aqueous medium with a yield 40%, are insoluble in water.

3. Synthesis and Modification of New Acrylonitrile Polymers and Copolymers with the Use of Polymer-Analogous Transformations of the Nitrile Group

The presence of nitrile groups in the PAN macromolecule having a high reactivity in nucleophilic and electrophilic reactions, opens many possibilities of conducting various chemical transformations using the reactions known for low-molecular nitriles. So far, only a very small number of these reactions has been used for the modification of PAN and AN copolymers. Besides this, the reactivity of the nitrile groups in the macromolecules of AN polymers and copolymers has not been studied thoroughly enough, and, in particular, the principal question on the difference of activity in the polymer molecules and the corresponding low-molecular analogues has not been answered.

3.1. Transformation of the Nitrile Groups of PAN into Aldehyde Groups and Derivatives

One of the interesting reactions of nitrile groups, which has been carried out rather recently on PAN, is their reduction according to Stephen which makes it possible to insert aldehyde groups into the macromolecule of PAN. It was found that in solutions of PAN in DMF saturated with dry hydrogen chloride and aged at temperatures of 0 to 5 °C a small quantity of the nitrile groups turns into imino chloride groups, which when treated with tin chloride, form aldimine stannates. These are converted by subsequent hydrolysis into aldehyde groups.

The reduction of PAN according to Stephen in dioxane at 60 to 100 °C made it possible, practically without any side reactions, to obtain modified PAN derivatives whose composition is close to that of copolymers from AN and acrolein.

$$\ldots-CH_2-CH-\ldots \xrightarrow{HCl} \ldots-CH_2-CH-\ldots \xrightarrow{HCl} \ldots-CH_2-CH-CH_2-CH-\ldots$$
$$\quad\quad | \quad\quad\quad\quad\quad\quad\quad\quad | \quad\quad\quad\quad\quad SnCl_2 \quad\quad\quad\quad | \quad\quad\quad\quad |$$
$$\quad\quad CN \quad\quad\quad\quad\quad\quad\quad C=\overset{\oplus}{N}H_2 \quad\quad\quad\quad\quad\quad\quad C=\overset{\oplus}{N}H_2 \quad CN$$
$$\quad\quad\quad\quad\quad\quad\quad\quad\quad\quad\quad\quad | \quad\quad \overset{\ominus}{Cl} \quad\quad\quad\quad\quad | \quad\quad HSnCl_4^{\ominus}$$
$$\quad\quad\quad\quad\quad\quad\quad\quad\quad\quad\quad\quad Cl \quad\quad\quad\quad\quad\quad\quad\quad Cl$$

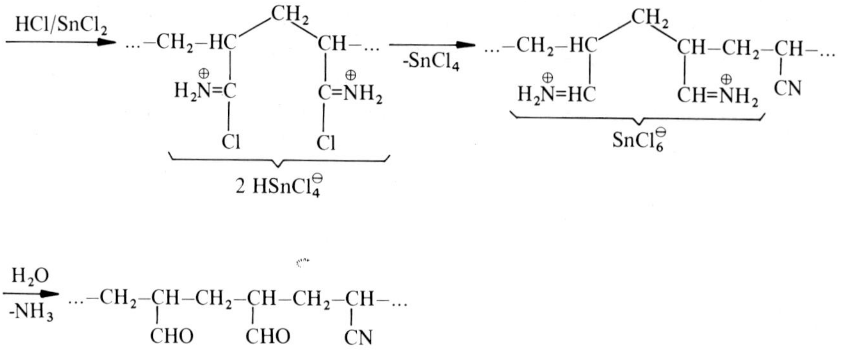

$$\xrightarrow{\text{HCl/SnCl}_2} \quad \underbrace{\begin{array}{c} \ldots-CH_2-HC \\ | \\ H_2\overset{\oplus}{N}=C \\ | \\ Cl \end{array} \begin{array}{c} CH_2 \\ \\ \end{array} \begin{array}{c} CH-\ldots \\ | \\ \overset{\oplus}{C}=NH_2 \\ | \\ Cl \end{array}}_{2\ HSnCl_4^{\ominus}} \xrightarrow{-SnCl_4} \underbrace{\begin{array}{c} \ldots-CH_2-HC \\ | \\ H_2\overset{\oplus}{N}=HC \end{array} \begin{array}{c} CH_2 \\ \\ \end{array} \begin{array}{c} CH-CH_2-CH-\ldots \\ | \\ \overset{\oplus}{CH}=NH_2 \quad CN \end{array}}_{SnCl_6^{\ominus}}$$

$$\xrightarrow[-NH_3]{H_2O} \ldots-CH_2-\underset{CHO}{CH}-CH_2-\underset{CHO}{CH}-CH_2-\underset{CN}{CH}-\ldots$$

This reaction was also applied to reduce the nitrile groups in the macromolecules of cellulose ethyl cyanates and poly(vinyl alcohol)[51]. In this case it has been found that along with aldehyde groups a considerable amount of carboxylic groups is also formed.

The presence of aldehyde groups in the macromolecules of modified PAN made it possible to obtain, as a result of an interaction with dimethyl phosphite at 120 °C, new PAN derivatives containing hydroxy-dimethoxyphosphorylmethyl groups according to the following scheme:

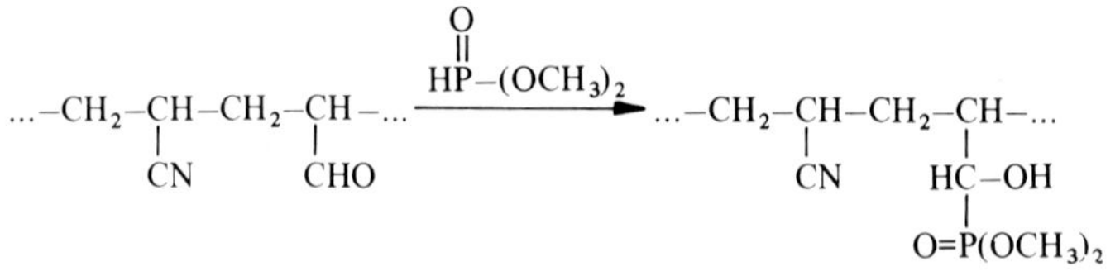

$$\ldots-CH_2-\underset{CN}{CH}-CH_2-\underset{CHO}{CH}-\ldots \xrightarrow{H\overset{O}{\overset{\|}{P}}-(OCH_3)_2} \ldots-CH_2-\underset{CN}{CH}-CH_2-\underset{\underset{O=P(OCH_3)_2}{|}}{\underset{HC-OH}{CH}}-\ldots$$

These phosphorus-containing PAN derivatives are unstable under hydrolytic conditions, and the phosphoryl groups are easily split off under the action of boiling water. If, however, the modified PAN is treated with dimethyl phosphite solution in toluene in the presence of dimethylamine, a modified PAN, stable towards hydrolysis, is obtained. Its composition seems to be the following:

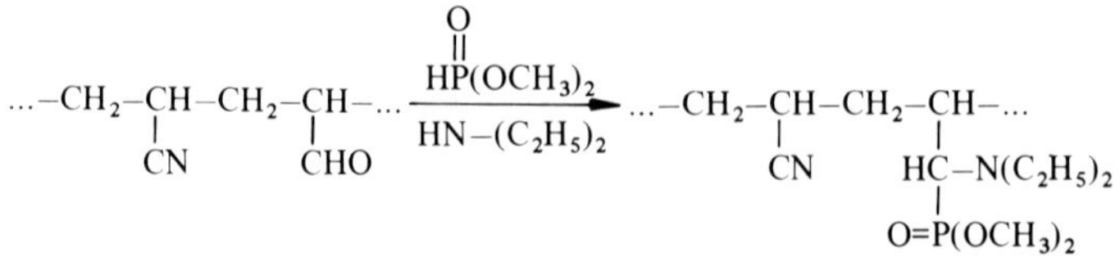

$$\ldots-CH_2-\underset{CN}{CH}-CH_2-\underset{CHO}{CH}-\ldots \xrightarrow[HN-(C_2H_5)_2]{H\overset{O}{\overset{\|}{P}}(OCH_3)_2} \ldots-CH_2-\underset{CN}{CH}-CH_2-\underset{\underset{O=P(OCH_3)_2}{|}}{\underset{HC-N(C_2H_5)_2}{CH}}-\ldots$$

Fibres phosphorylated in this way and containing 8–10% of diethylamino-dimethoxyphosoryl groups are incombustible.

3.2. Transformation of the Nitrile Groups of PAN into Amino Groups; Thioamidation

Investigations of PAN reduction in DMF solution in the presence of Raney nickel, sodium hyposulfite, acetic acid, pyridine, and water at 40–100 °C have shown[52] that, in contrast to the quantitative reduction of low-molecular nitriles under the same conditions to the corresponding aldehydes[53], the reduction of the nitrile groups of PAN proceeds primarily in the direction of the formation of amine groups (20%), and only of an insignificant (1–2%) quantity of aldehyde groups. At the same time, seemingly as a result of the intramolecular interaction of the formed amino groups with the adjacent nitrile groups, cyclization is taking place and piperidine units (2–3%) are formed. This reaction can be presented as follows:

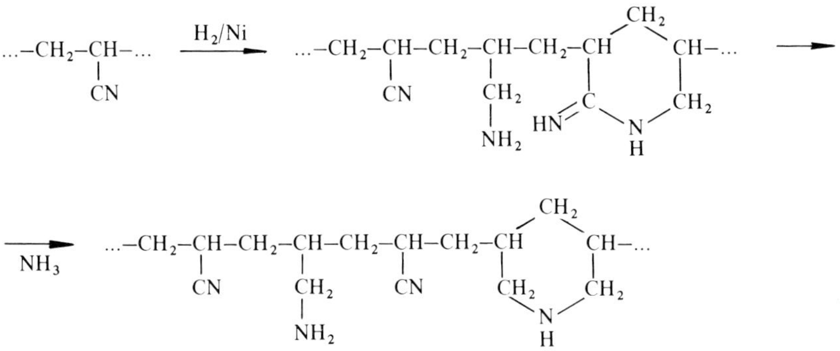

$$\ldots-CH_2-CH-\ldots \xrightarrow{H_2/Ni} \ldots-CH_2-CH-CH_2-CH-CH_2-CH \diagdown CH-\ldots \longrightarrow$$
$$||||CH_2$$
$$CNCNCH_2C\diagupCH_2$$

$$\xrightarrow{NH_3} \ldots-CH_2-CH-CH_2-CH-CH_2-CH-CH_2-CH \diagdown CH-\ldots$$

It has been shown[52] that under similar conditions reduction of the nitrile groups in cellulose ethyl cyanate and of those in the copolymer of vinylidene cyanide with vinyl acetate, proceed simultaneously in two directions with the formation of aldehyde and amine groups.

The nitrile groups, $-\overset{\delta+}{C}\equiv\overset{\delta-}{N}$, are strongly polarized and, as a result, the carbon atom becomes positively charged, which predetermines the high reactivity of the nitrile group in electrophilic reactions when interacting with nucleophilic reagents and, in particular, with hydrogen sulfide.

The reaction of PAN with hydrogen sulfide was for the first time achieved in DMF solution at 50–70 °C. It has been shown that under these conditions the nitrile groups are partially converted into thioamide groups and polymers corresponding to copolymers of AN with acrylthioamide are formed[54]:

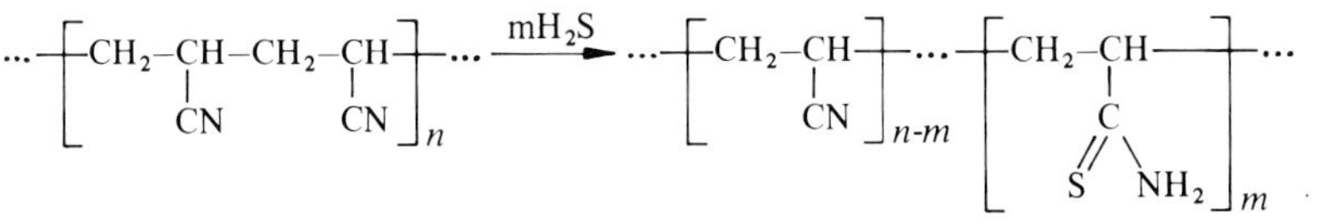

$$\ldots\left[CH_2-CH-CH_2-CH\right]_n\ldots \xrightarrow{mH_2S} \ldots\left[CH_2-CH\right]_{n-m}\left[CH_2-CH\right]_m\ldots$$

This reaction has recently been applied with success[55] for the modification of poly-(α-chloroacrylonitrile). As reported by the authors, the level of conversion of the nitrile groups into thioamide groups reaches 90%.

The necessary condition to assure thioamidation of PAN is the use of basic catalysts, and, in particular, of amines facilitating the formation of the mercapto anion in the reaction medium.

This fact gives enough reason to suppose that the reaction of hydrogen sulfide addition to nitrile groups starts with the attack of the nucleophilic mercapto anion at the positively charged carbon atom of the nitrile group, and then the reaction is accomplished due to the ease of the proton transfer to the nitrogen atom.

The reaction mechanism can be presented as follows[56]:

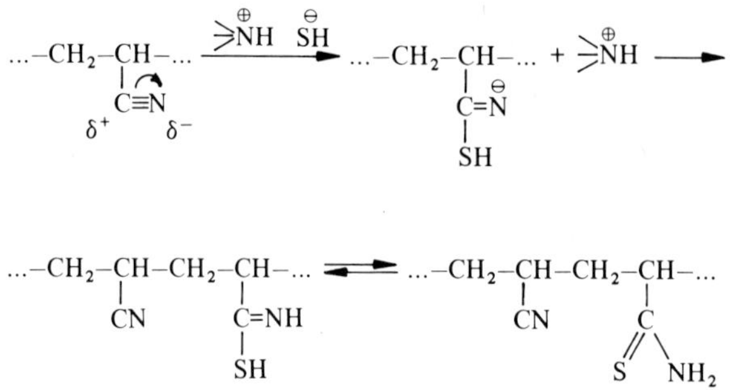

$$...-CH_2-CH-... \xrightarrow{\overset{\oplus}{>}NH \ \overset{\ominus}{SH}} ...-CH_2-CH-... + \overset{\oplus}{>}NH \longrightarrow$$
$$\underset{\delta^+}{|} \underset{\delta^-}{C\equiv N} \qquad \qquad \underset{SH}{\overset{|}{\underset{|}{C=N}}^{\ominus}}$$

$$...-CH_2-CH-CH_2-CH-... \rightleftharpoons ...-CH_2-CH-CH_2-CH-...$$
$$\qquad \ \ |\qquad \qquad |\qquad \qquad \qquad \ \ |\qquad \qquad |$$
$$\qquad \ CN \qquad \ \ C=NH \qquad \qquad \ CN \qquad \ \ C$$
$$\qquad \qquad \qquad \quad \ \ | \qquad \qquad \qquad \qquad \qquad //\backslash$$
$$\qquad \qquad \qquad \ \ SH \qquad \qquad \qquad \qquad \quad S \quad NH_2$$

$$R_3N + H_2S \longrightarrow R_3\overset{\oplus}{N}H \ \overset{\ominus}{SH}$$

It was found later[57] that thioamidation of PAN proceeds readily and practically without any side reaction even at 0–5 °C when ammonium hydrosulfide is used as the reagent for the thioamidation.

The reaction of the thioamidation of PAN is characterized by the following distinguishing features:

1. Thioamidation with hydrogen sulfide at low temperature (10–20 °C) in the presence of a small quantity of NH_4SH, used as catalyst, proceeds only up to 50%. Under similar conditions lowmolecular model compounds of PAN-acetonitrile and glutarodinitrile-correspondingly form thioamides with high yields. It has been shown that in the thioamidation of 1, 3, 5-pentane-tricarbonitrile only one group turns into a thioamide group. This seems to be taking place in accordance with the following scheme:

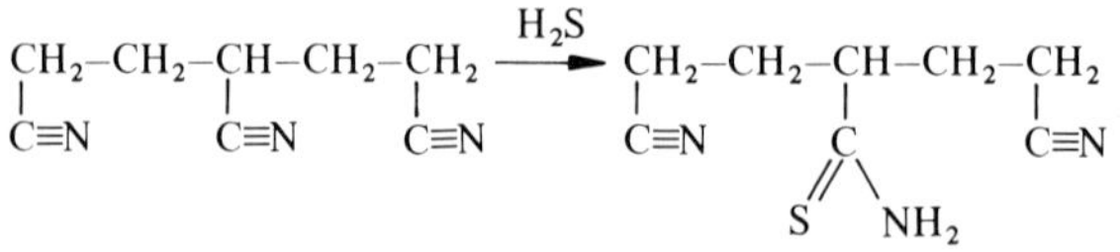

$$CH_2-CH_2-CH-CH_2-CH_2 \xrightarrow{H_2S} CH_2-CH_2-CH-CH_2-CH_2$$
$$|\qquad \qquad |\qquad \qquad |\qquad \qquad \qquad |\qquad \qquad |\qquad \qquad |$$
$$C\equiv N \qquad C\equiv N \qquad C\equiv N \qquad \qquad C\equiv N \qquad C \qquad C\equiv N$$
$$\qquad \qquad \qquad \qquad \qquad \qquad \qquad \qquad \qquad \qquad \qquad //\backslash$$
$$\qquad \qquad \qquad \qquad \qquad \qquad \qquad \qquad \qquad \qquad S \quad NH_2$$

This makes it possible to assume that by thioamidation of PAN under mild conditions, modified polymers are obtained composed of macromolecules with alternating nitrile and thioamide groups[56].

2. In thioamidation the nitrile groups of PAN have a much higher reactivity than those of the corresponding model compounds. This fact is explained by the specific character of the polymeric nature of PAN and by the mutual influence of adjacent groups. As it is seen from the data presented in Fig. 4, the highest reaction rate and conversion level, as compared with low-molecular nitriles, is observed in the thioamidation of PAN.

This fact can be explained by the mutual effect of the nitrile groups in the macromolecules of the PAN. It seems reasonable to assume that the nitrile group,

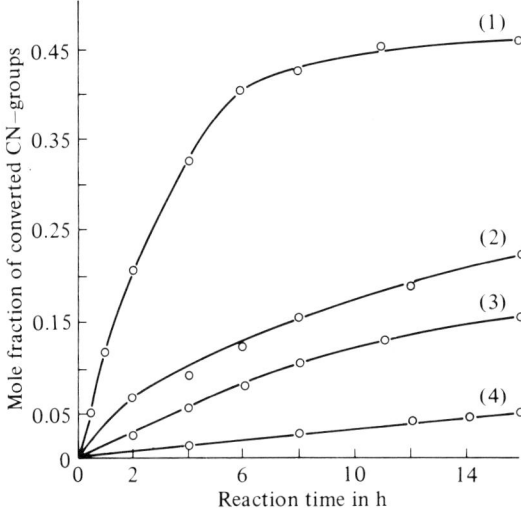

Fig. 4. Dependence of the conversion level on the time of thioamidation of PAN and low-molecular nitriles at 20 °C in DMFA. (1): PAN: (2): 1,35-pentanetricarbonitrile. (3): glutarodinitrile. (4): acetonitrile

due to its nucleophilic properties determined by the presence of the electron-donating π-bond and the free electron pair, interacts with the positively charged carbon atom of the adjacent nitrile groups, contributing in this way a shift of the electron density in one of the nitrile groups and to an increase of the positive charge at the carbon atom:

This, naturally, results in an increased electrophilicity of the nitrile group and also creates favourable conditions for the nucleophilic attack of the mercapto anion and an easy addition of hydrogen sulfide in accordance with the above scheme. The activation energy of the thioamidation of the model compounds is much higher (for glutarodinitrile − 11,8 [49,4] and for trinitrile − 7,97 kcal/mol [33,4 kJ/mol]) than in PAN (6,18 kcal/mol [25,9 kJ/mol]).

The highest value of activation energy is observed in acetonitrile (17,52 kcal/mol [73,35 kJ/mol]), in which the mutual effect of the nitrile groups is absent.

The mutual effect of the adjacent groups on the reactivity of the nitrile group in the thioamidation reaction manifests itself to an even greater extent when hydrogen sulfide interacts with AN copolymers[58]. As it is seen from the data in Fig. 5, the rate and level of the conversion of nitrile groups into thioamide groups is much higher for copolymers of AN with acrolic acid (AA), with methyl acrylate (MA),

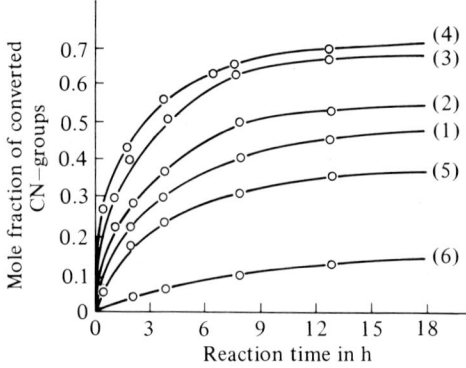

Fig. 5. Effect of the nature of adjacent functional groups in copolymers on the rate of conversion of the nitrile groups into thioamide groups. (1): PAN: (2): AN-VDC: (3): AN-AA: (4): AN-MA: (5): AN-VC: (6): AN-ST

with vinyl acetate (VA) or with vinylidene chloride (VDC) than in the case of PAN, which is indicative of a comparatively higher reactivity of the nitrile groups in the macromolecules of the copolymers.

The observed differences in the reactivity of the nitrile groups in the macromolecules of the copolymers can be explained by an intramolecular nucleophilic interaction of the nitrile groups with the adjacent functional groups applying the same point of view as for adjacent nitrile groups:

Due to the fact that the nitrile groups interact with the positively charged carbon atoms of the carboxyl or ester groups more easily than the less mobile nitrile groups in the PAN macromolecules interact with each other, the electrophilicity of the nitrile groups in the macromolecules of the copolymers increases to a greater extent, which, naturally, manifests itself in the increase of the rate of hydrogen sulfide addition.

Insertion of styrene units into the macromolecules of AN polymers causing steric hindrances because of the threedimensional phenyl groups, results in an appreciable decrease of the rate of thioamidation in comparison with PAN.

The reaction of hydrogen sulfide with PAN or with AN copolymers proceeds in an interesting and peculiar way when they are treated in a heterogeneous medium with aqueous solutions of ammonium sulfide at 80–100 °C[60]. Under these conditions, simultaneously with the thioamidation, various side reactions are also taking place to a large extent, their nature and rate being significantly dependent on the nature of the adjacent groups. The shape of the curves showing the level of conversion of the nitrile groups into thioamide groups vs. the reaction time, which are given in Fig. 6, indicates that in the case of thioamidation of PAN or of the copolymers of AN containing carboxylic or ester groups, the thioamide groups formed in the course of the reaction participate in side reactions.

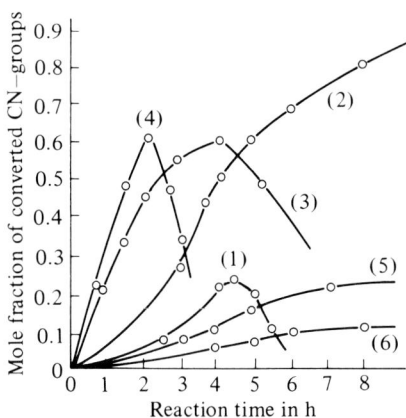

Fig. 6. Effect of the nature of adjacent functional groups in copolymers on the level of conversion of nitrile groups. (1); PAN; (2); AN-AA: (3); AN-MA; (4); AN-VDC; (5); AN-VC; (6); AN-ST

An investigation of the thioamidation reaction on model compounds of PAN under similar conditions and a study of the functional composition of modified copolymers led to the assumption that the most probable course of the two types of side reactions causing the decrease of the number of thioamide groups are hydrolysis and cyclization reactions.

1. Hydrolysis:

$$\ldots-CH_2-CH-\ldots \xrightarrow[-H_2S]{H_2O} \ldots-CH_2-CH-\ldots$$
$$\underset{S}{\overset{C}{\|}}{-NH_2} \qquad \underset{O}{\overset{C}{\|}}{-NH_2}$$

2. Intramolecular cyclization with subsequent hydrolysis. (See page 26.)
3. Along with these reactions cyclization of adjacent nitrile groups is also taking place with the formation of conjugated naphthyridine rings. This is the most important and valuable feature of thioamidation of PAN since it makes it possible to obtain modified PAN fibres with a staircase structure and exceptionally high heat resistance under comparatively simple conditions[58].

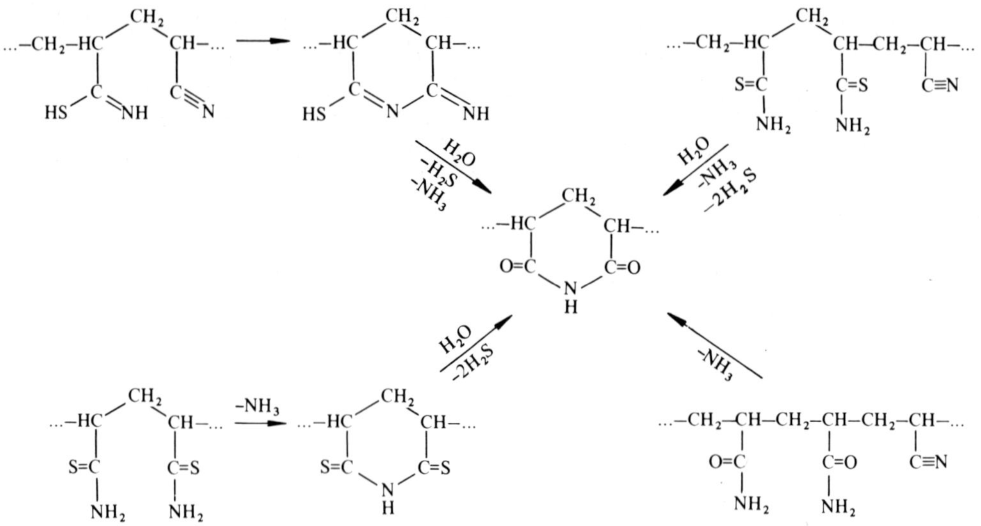

It has been established[59] that cyclization is initiated by the presence of thioamide groups in the macromolecules of modified PAN. This is confirmed by the fact that, when PAN is treated with ammonium hydroxide, nitrile groups do not undergo any transformations, whereas in modified PAN containing even a small amount (3–5%) of thioamide groups this treatment results in a significant change of its functional composition, *i.e.* absorption bands characteristic of nitrile groups disappear almost completely from the IR spectra and new bands characteristic of conjugated —C=N— bonds appear. At the same time, the colour of the sample changes to dark orange, and the exothermic effect caused by cyclization of the nitrile groups decreases as shown by DTA curves. The mechanism of cyclization initiation resulting in the formation of naphthyridine cycles with the participation of thioamide groups can be illustrated by the following scheme:

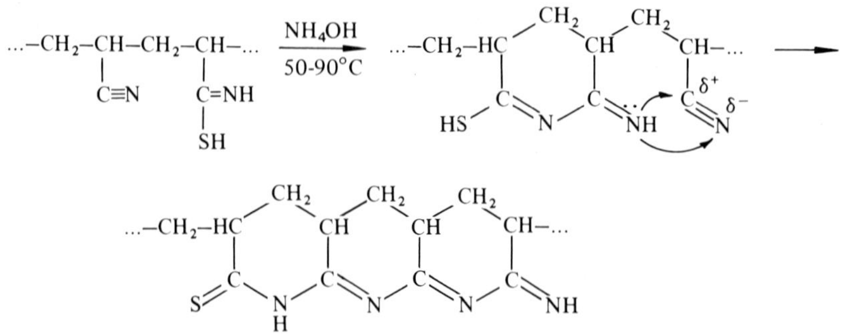

The formation of such a staircase structure in the macromolecules of PAN in the process of thioamidation of PAN fibres with aqueous solutions of ammonium sulfide accounts for the significant increase of heat resistance of the modified fibres obtained.

At 180–200 °C these fibres preserve 60–70% of their initial strength, whereas PAN fibres under these conditions practically lose their strength completely.

An even higher heat resistance of modified AN copolymers is attained as a result of cyclization by treating previously crosslinked fibres based on the copolymer of AN and 4-chloro-2-butenyl methacrylate[44] with ammonium sulfide:

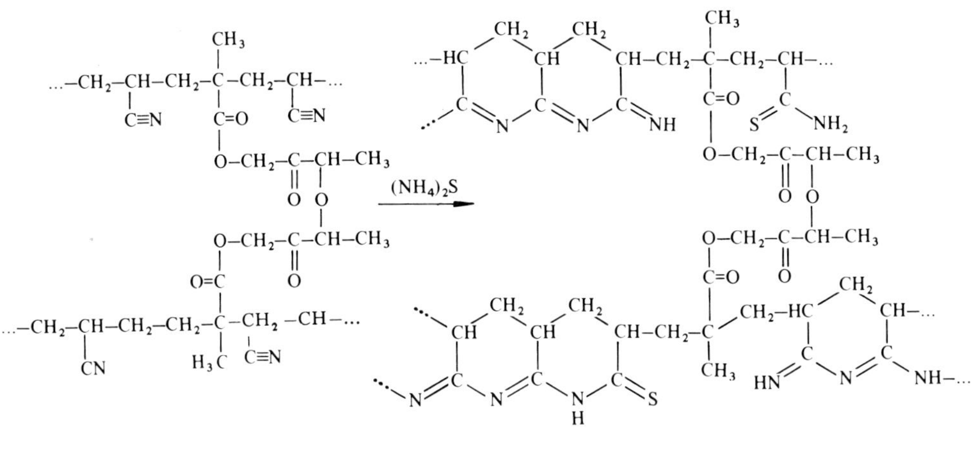

Combining the processes of crosslinking and cyclization provides the possibility of obtaining modified PAN fibres which have a very high heat resistance and preserve 95–97% of their strength at 180–200 °C.

Other types of side reactions take place with the thioamidation of copolymers from AN and vinylidene chloride or vinyl chloride[60].

These copolymers undergo side reactions which are mainly reducible to reactions in which chloride atoms take part. When copolymers of AN and vinylidene chloride are treated with aqueous solutions of ammonium sulfide at 90 °C, nitrile groups have been shown to be almost fully converted into thioamide groups, and practically no hydrolysis of the formed thioamide groups is taking place in this case. At the same time, it was found that in the course of this treatment dehydrochlorination is taking place almost quantitatively:

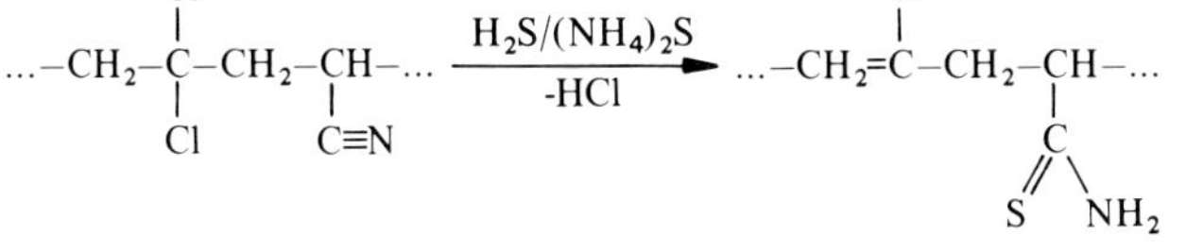

The process of thioamidation of the copolymers of AN with vinyl chloride has an even more complicated nature. As it has been shown[60], in this case no dehydrochlorination is taking place and not only the nitrile groups react with hydrogen sulfide but the chlorine atoms as well. However, even at stages of high conversion, unreacted nitrile groups were found to remain. Taking into account the experimental results of the reaction of poly(vinylchloride) with hydrogen sulfide[61], the process of thioamidation of copolymers of AN with vinyl chloride can in its general form be visualized as formulated in the following scheme

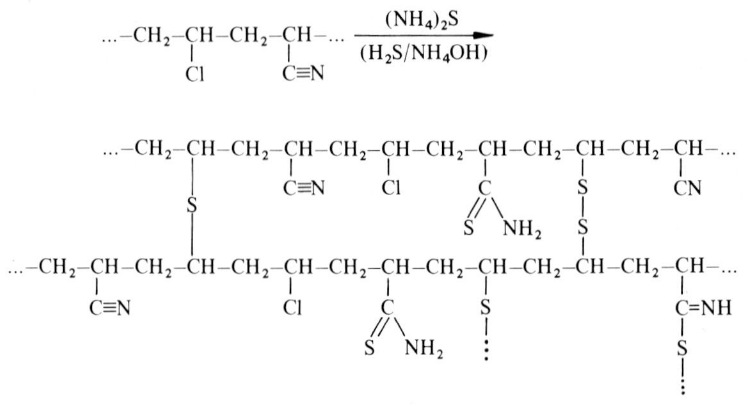

$$...-CH_2-CH-CH_2-CH-... \xrightarrow[(H_2S/NH_4OH)]{(NH_4)_2S}$$
$$||$$
$$ClC\equiv N$$

[Scheme showing two parallel polymer chains with side groups: $C\equiv N$, Cl, $C(=S)NH_2$, S, CN on upper chain; and $C\equiv N$, Cl, $C(=S)NH_2$, S, $C=NH$ on lower chain, with sulfur bridges between the chains]

Though no naphthyridine rings are formed when copolymers of AN with vinylidene chloride (50:50) or with vinyl chloride (50:50) are treated with ammonium sulfide solutions, it is, however, probable that, as a result of a considerable number of intermolecular chemical bonds being formed in the macromolecules or copolymers, the heat resistance of these modified copolymers

Table 3. Properties of thioamidated fibres

Fibre[a]	Mechanical properties		Elastic modulus in kgf/mm^2		Heat resistance. Preservation of strength		Breaking temp. in °C[b]	Thermal stability in %[c]
	strength in $\frac{kgf}{tex}$	elongation in %	at 20 °C	at 200 °C	at 140 °C	at 200 °C		
PAN	35–38	16–20	580	25	20–30	5–8	137–140	71–75
PAN-T	30–35	21–28	560	300	60–70	55–60	290–310	60–62
Crotylan	30–32	9–11	215	10	25–28	5–7	130	60–65
Crotylan CL	28–29	10–12	275	20	29–30	20–22	180	65–70
Crotylan T	30–32	8–10	230	260	90–92	75–80	280	–
Crotylan CLT	30–31	8–10	280	450	100	95–97	330	–
Caniv	39,1	11,0	630	0	5–10	0	90–95	0
Dainel	30,5	10,0	590	0	8–10	0	70–75	0
Caniv-T	35,0	11,0	630	440	70–75	45–50	360	35–40
Dainel-T	29,5	9,5	580	300	70–75	55–58	240	40–44

[a] T after the name of the fibre stands for "thioamidated". CL-fibre previously crosslinked with ZnO. Crotylan – fibre from AN copolymer with 3-chloro-2-butenyl methacrylate. Caniv – fibre from AN copolymer with vinylidene chloride.
[b] At a load to 10% of the tensile strenght.
[c] Preservation of strength after 5 h heating at 200 °C.

also sharply increases[62]. In the process of thioamidation of fibres obtained from these copolymers a considerable decrease of chlorine content is observed, but nevertheless, they remain incombustible[60].

Some of the properties of fibres obtained from PAN and from AN copolymers modified with acqueous solutions of ammonium sulfide are shown in Table 3.

An interesting feature of modified PAN containing thioamide groups (PAN-T) is that the presence of even a small amount ($\approx 5\%$) of thioamide groups in the macromolecules of PAN initiates the thermal cyclization and lowers the temperature at which its cyclization starts down to 130–150 °C[59].

Results of investigation of the process of thermal cyclization of modified PAN showed thioamide groups to be unstable at elevated temperatures, and, when PAN-T is heated, hydrogen sulfide evolves as a result of the decomposition of thioamide groups. It has been shown that heating of the polymer directly in the heating zone at a given temperature (110–170 °C) for 30–90 min, results in an intensive decomposition of the thioamide groups with complete removal of sulfur in the form of hydrogen sulfide. If the thermal treatment of PAN-T is carried out with a gradual increase of temperature in the heated sample, a certain amount of sulfur remains. In both cases this process is accompanied with the formation of conjugated –C=N– bonds, which is indicative of the participation of the thioamide groups in the cyclization reaction.

The following scheme can be suggested for the initiation of PAN cyclization by thioamide groups:

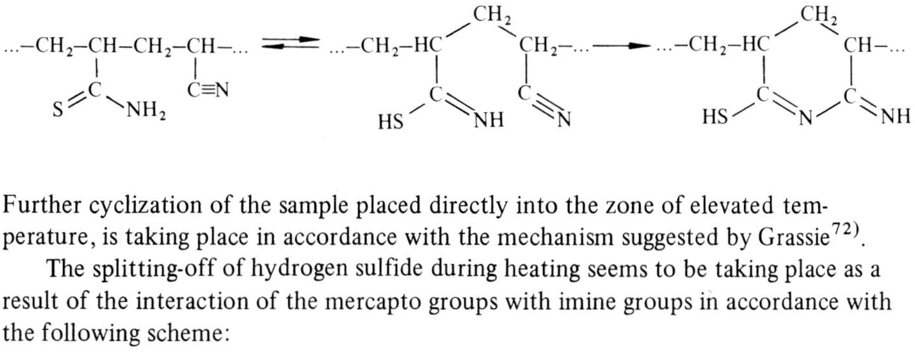

Further cyclization of the sample placed directly into the zone of elevated temperature, is taking place in accordance with the mechanism suggested by Grassie[72].

The splitting-off of hydrogen sulfide during heating seems to be taking place as a result of the interaction of the mercapto groups with imine groups in accordance with the following scheme:

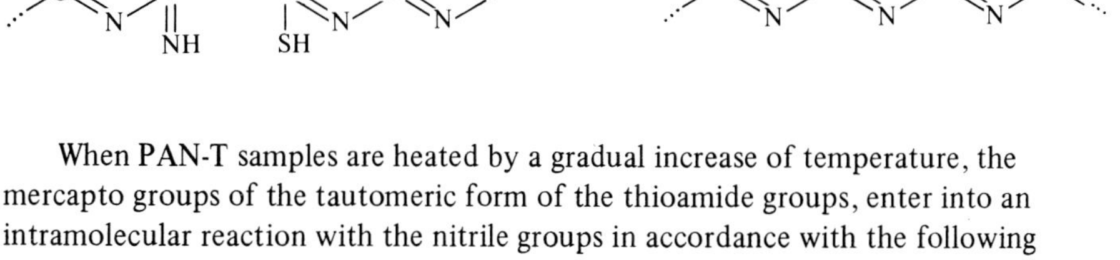

When PAN-T samples are heated by a gradual increase of temperature, the mercapto groups of the tautomeric form of the thioamide groups, enter into an intramolecular reaction with the nitrile groups in accordance with the following scheme:

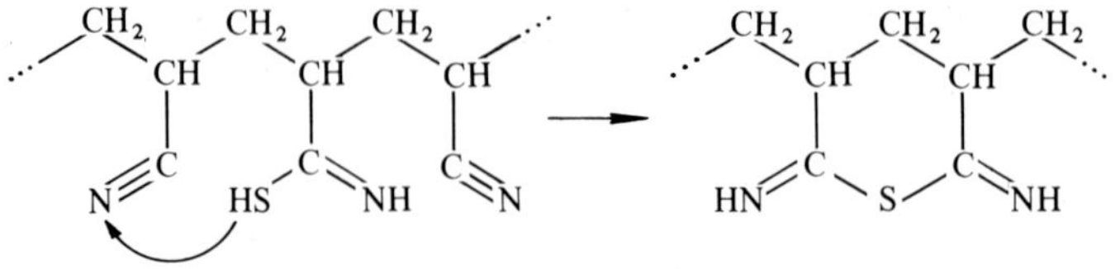

$$\underset{N}{\overset{CH_2}{\underset{|}{\text{CH}}}}\underset{HS}{\overset{CH_2}{\underset{|}{\text{CH}}}}\underset{NH}{\overset{CH_2}{\underset{|}{\text{CH}}}}\cdots \longrightarrow \cdots \underset{HN}{\overset{CH_2}{\underset{|}{\text{CH}}}}\underset{S}{\overset{CH_2}{\underset{|}{\text{CH}}}}\overset{CH_2}{\underset{NH}{\underset{|}{\text{CH}}}}\cdots$$

The imine group formed as a result of this reaction will then initiate the reaction of cyclization proceeding in accordance with the above scheme.

The activation energy of the reaction of thermal cyclization of PAN containing 5% of thioamide groups amounts to 20 kcal/mol (83,7 kJ/mol).

4. Synthesis of Block and Graft Copolymers of Polyacrylonitrile

Recent achievements in the field of producing elastomers, of some kinds of plastics, and of modifying cellulose fibres by the synthesis of block and graft copolymers show these methods to hold great promise for the purposeful modification of polymer properties and the creation of materials with improved or new properties. In a number of cases synthesis of block and graft copolymers is the most expedient method of obtaining fibres having special properties. Therefore, in the recent years research on the modification of fibre-forming polymers and on fibres obtained from them by synthesizing block and graft copolymers, has found an ever growing development.

Significant progress has been made in the field of synthesizing graft copolymers of cellulose[62], polyamide[63], and poly(vinyl alcohol)[64]. At the same time, the development of acceptable and practicable methods for the synthesis of PAN graft copolymers has been neglected. PAN is known to contain no reactive functional groups that would make it possible to conduct the reaction of graft copolymerization using chemical methods of initiation. Therefore, in most cases, to obtain graft copolymers of PAN, copolymers of AN containing functional groups with reductive properties, etc. are used.

4.1. Graft Copolymers of PAN *via* Diazonium Salts

In this respect AN copolymers with 2-amino-4-vinylsulfonylanisole 7 are of considerable interest. The presence of a small quantity (1–2%) of monomeric units of 7 containing aromatic amino groups in the copolymer macromolecules makes it possible to apply two methods for the synthesis of PAN graft copolymers[65, 66]. Following one of them[65] the grafting is conducted by diazotizing the amino group followed by the decomposition of the diazo groups in the presence of Fe^{2+}. The macroradicals, thus formed, initiate the reaction of graft copolymerization. This method of grafting does not produce any homopolymer. The reaction can be formulated as follows:

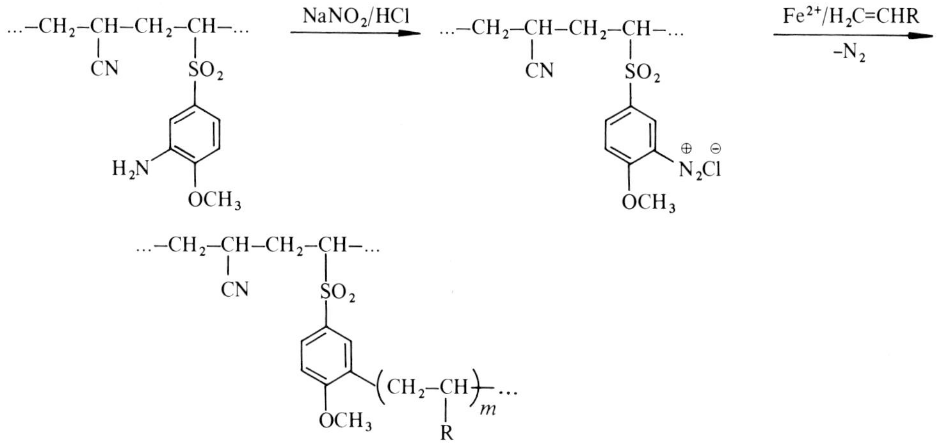

4.2. Graft Copolymers of PAN with the Use of Redox Systems

The other method[66] which also makes it possible to obtain graft copolymers with a number of vinyl monomers practically without any homopolymer being formed, is based on using the oxidation-reduction systems in which the copolymer of AN with 7 plays the part of the reducing agent and pentavalent vanadium compounds, e.g., vanadic acid, are the oxidizing agents. Vanadic acid is known to dissociate in acid medium as follows:

$$HVO_3 \rightleftharpoons VO_2^+ + OH^-$$

As a result of the interaction of VO_2^+ ion with the amino groups of the copolymer macroradicals are formed which initiate the reaction of graft copolymerization. The mechanism of this reaction is illustrated by the following scheme:

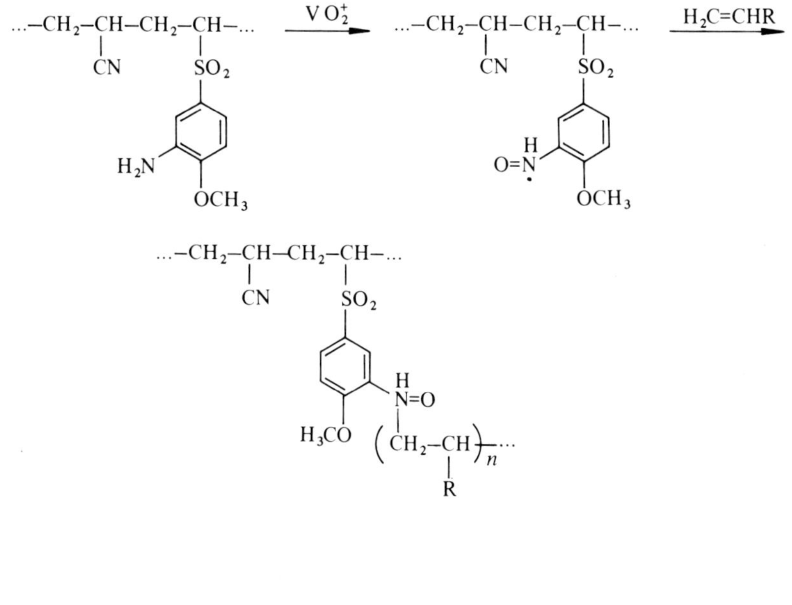

By this method graft copolymers of PAN with poly(methyl acrylate) (PMA), poly-(butyl acrylate) (PBA), poly(acrylic acid) (PAA), poly(methyl vinylpyridine) (PMCP), and polystyrene (PSI) have been obtained.

Systematic studies[67] to determine the properties of solutions of the obtained graft copolymers showed that graft copolymers containing stiff main chains (PAA, PMVP, PST) dissolve in DMF only at elevated temperatures and when the content of grafted component is very small (1–3%), in spite of DMF being a solvent to all of them. Graft copolymers containing flexible side chains (PMA, PBA) are dissolved, but their solubility decreases with increasing degree of grafting. Fibres spun from graft copolymers have a better combination of physico-mechanical parameters than that of modified fibres obtained by grafting to the finished fibre made of AN copolymers with 7. It seems that in the first case a more perfect super-molecular structure is formed.

To synthesize graft copolymers of PAN using oxidation-reduction systems, random copolymers of AN with methacrolein (MAC) can also be taken as the initial product. In this case, aldehyde groups of the copolymer macromolecule serve as the reducing agent and cerium (Ce^{4+}) salts as the oxidizing agent.

Since Ce^{4+} salts are capable of causing the homopolymerization of vinyl monomers starting after a certain induction period, the grafting process is carried out during a time period shorter than the period of induction so as to synthesize graft PAN copolymers without any homopolymer being formed[68].

The mechanism of the reaction of graft copolymerization, using copolymers of AN-MAC and Ce^{4+} salts, can be illustrated by the following scheme:

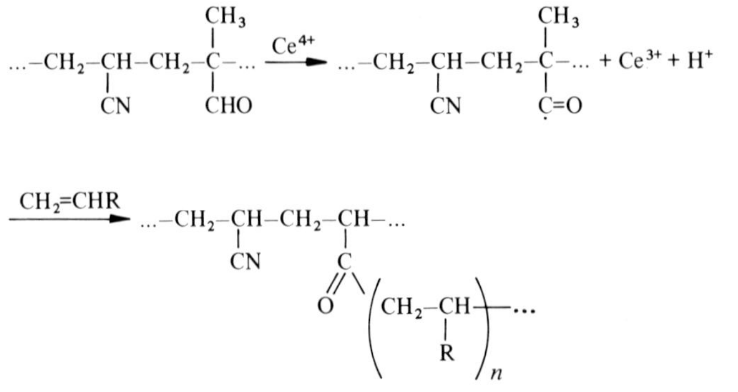

$$...-CH_2-CH(CN)-CH_2-C(CH_3)(CHO)-... \xrightarrow{Ce^{4+}} ...-CH_2-CH(CN)-CH_2-\overset{\cdot}{C}(CH_3)(C=O)-... + Ce^{3+} + H^+$$

$$\xrightarrow{CH_2=CHR} ...-CH_2-CH(CN)-CH_2-CH\left(\underset{O}{\overset{C}{\underset{\parallel}{\diagdown}}}(CH_2-CH(R))_n\right)-...$$

Graft copolymers of PAN with PMA have been synthesized by this method and branched PAN was obtained using AN as the grafted monomer. Fibres obtained from graft copolymers, irrespective of the nature of the grafted component, are superior to PAN fibres as far as their resistance to multiple deformations is concerned.

A new method of synthesizing graft PAN copolymers with poly(acrylic acid) has recently been proposed. It is based on using an oxidation-reduction system in which modified PAN containing 7–10% of thioamide groups serves as the reducing agent and hydrogen peroxide as the oxidizing agent[69].

The initiation of graft copolymerization using this redox system, seems to proceed in accordance with the following scheme:

1. Thioamide tautomerism:

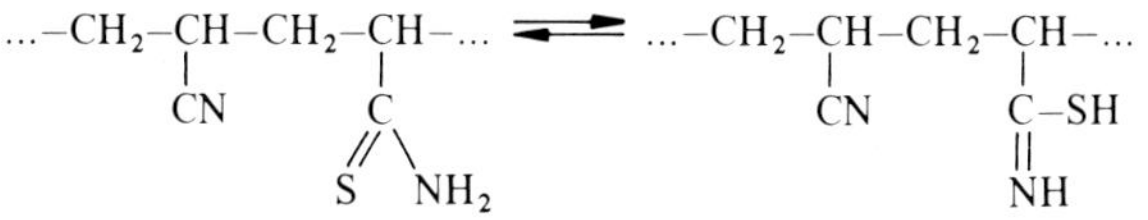

$$...-CH_2-CH-CH_2-CH-... \rightleftharpoons ...-CH_2-CH-CH_2-CH-...$$

with CN and C(=S)(NH$_2$) groups on the left side, and CN and C(-SH)=NH groups on the right side.

2. Initiation:

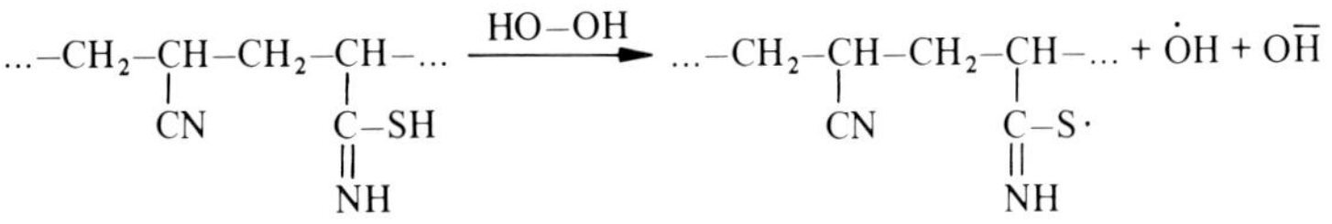

$$...-CH_2-CH-CH_2-CH-... \xrightarrow{HO-OH} ...-CH_2-CH-CH_2-CH-... + \dot{O}H + \overline{OH}$$

with CN and C(-SH)=NH on the left, CN and C(-S·)=NH on the right.

3. Chain propagation:

$$...-CH_2-CH-CH_2-CH-... + CH_2=CH \longrightarrow ...-CH_2-CH-CH_2-CH-...$$

with CN and C(-S·)=NH on the left reactant, CH$_2$=CH with C(=O)(OH) substituent as monomer, and the product having CN and $C(-S-[CH_2-CH(C(=O)OH)]_n)=NH$ chain.

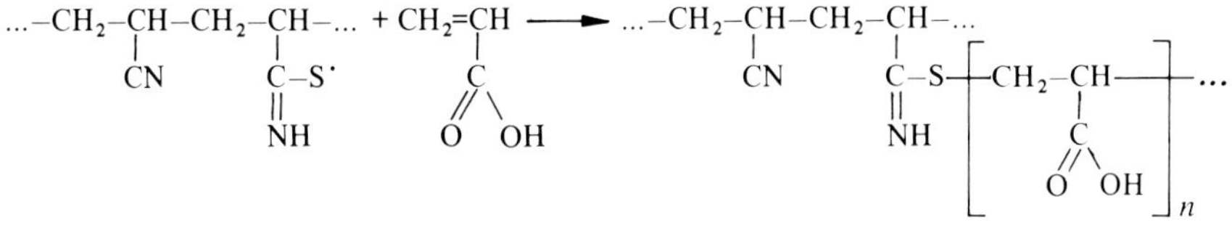

4. The chain ends as a result of the interaction between the growing macroradical and the radicals of the initiator

The investigations have shown, however, that graft copolymerization carried out according to this method is accompanied with a simultaneous reaction of monomer homopolymerization which, naturally, reduces the effectiveness of the method. This is explained by the presence of hydroxyl radicals in the reaction medium, which are formed as formulated in the above scheme.

4.3. Graft Copolymers of PAN with the Use of Oligomeric Diisocyanates

An interesting and in principle new method for the synthesis of graft PAN copolymers, which makes it possible to conduct the grafting process directly in the spinning solution, is the method based on the reaction of AN copolymers containing a small quantity of carboxylic groups with diisocyanates of relative high molecular weight[70].

The formation of graft copolymers of PAN with oligomeric diisocyanates proceeds in accordance with the following scheme:

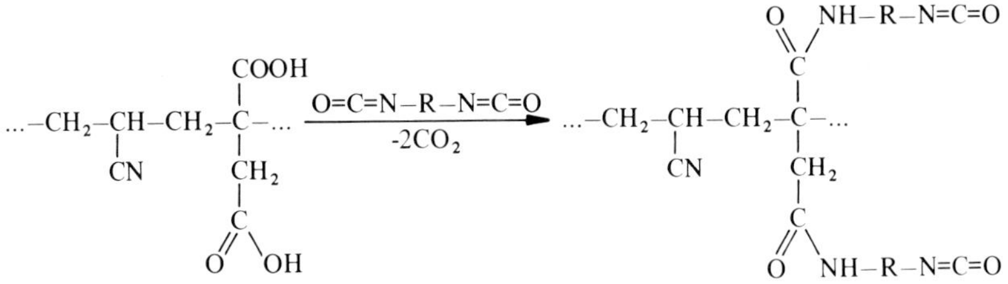

$$...-CH_2-CH-CH_2-C(COOH)- \xrightarrow[{-2CO_2}]{O=C=N-R-N=C=O} ...-CH_2-CH-CH_2-C-...$$

Left structure: CN substituent and CH$_2$-C(=O)OH substituent.
Right structure: CN substituent, CH$_2$-C(=O)(NH-R-N=C=O) and C(=O)(NH-R-N=C=O) substituents.

This method of synthesizing graft copolymers has a number of advantages, in particular:
1. The possibility of producing graft polymers with the same pre-assigned length of grafted chains;
2. the possibility of controlling the number of grafted chains, and,
3. the possibility of conducting various subsequent transformations, under mild conditions because of the presence of reactive isocyanate groups at the end of the grafted chains.

This makes it possible to produce chemically stained modified PAN fibres following the scheme below.

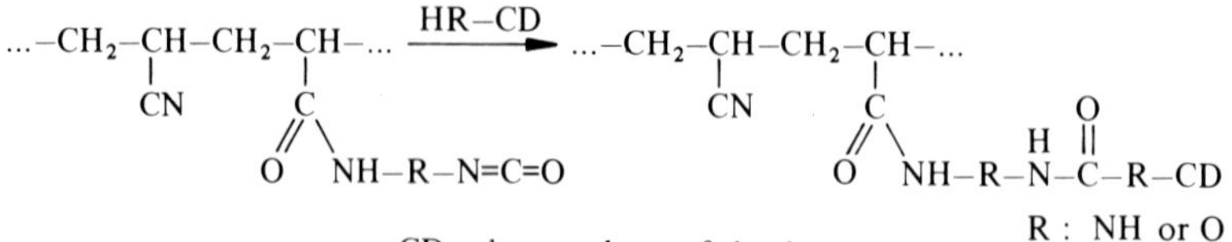

$$\ldots-CH_2-CH-CH_2-CH-\ldots \xrightarrow{HR-CD} \ldots-CH_2-CH-CH_2-CH-\ldots$$

with substituents: CN, C(=O)−NH−R−N=C=O → CN, C(=O)−NH−R−N(H)−C(=O)−R−CD

R : NH or O

CD — chromophore of the dye

A fundamental advantage of this version of the synthesis of graft PAN copolymers, as compared with the grafting to the finished fibre, is the fact that the grafted chains take part in forming the supermolecular structure of the fibre, which results in a considerabel improvement of the properties of modified fibres. Thus, fibres spun from the graft copolymer of PAN with poly(propylene oxide) (molecular mass $\approx$ 1000) have an abrasive resistance which is 5–6 times higher than that of the fibres obtained from a random copolymer. Fibres obtained from the solution of mixtures of AN copolymer with itaconic acid and poly(propylene oxide) have almost the same abovementioned parameters as those of the random copolymer[71].

4.4. Block Copolymers of PAN

For the modification of PAN properties the methods of synthesizing block copolymers of PAN can be of substantial interest.

Although the synthesis of block copolymers is a promising means of PAN modification with the purpose of producing fibres with enhanced performance characteristics, up to now very few investigations have been carried out to develop methods for the synthesis of PAN block copolymers. Meanwhile, a combination of stiff blocks consisting of AN units with flexible blocks having diverse chemical properties, provides the necessary prerequisites for varying within a broad range the physico- chemical and physico-mechanical properties of modified copolymers. In this respect, block copolymers of PAN with poly(ethylene oxide) (PEO) or poly-(propylene oxide) (PPO) are of interest.

As it was shown in[73, 74], methods that can be used to synthesize these copolymers of PAN are those of radical AN block copolymerization in the presence of an oxidation-reduction system in which the hydroxyl end groups of poly(ethylene oxide) (PEO)[73] and poly(propylene oxide) (PPO)[74] oligomers serve as the reducing agents and tetravalent cerium salts as the oxidizing agents.

When PAN block copolymers are synthesized with the help of such method the reaction is conducted during a period of time (20–30 min) shorter than the induction period of AN polymerization (45 min) in the presence of cerium ions. When the mechanism and the laws governing the reaction of AN copolymerization with PEO and PPO were studied, it was established that the initiation of the block copolymerization proceeds in accordance with the following scheme:

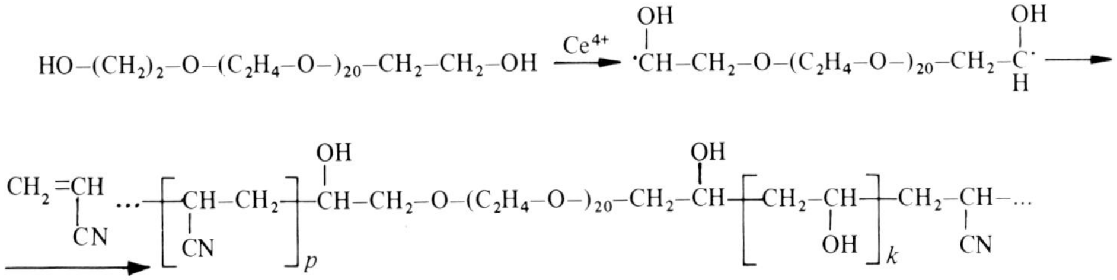

$$HO-(CH_2)_2-O-(C_2H_4-O-)_{20}-CH_2-CH_2-OH \xrightarrow{Ce^{4+}} \overset{OH}{\underset{}{\cdot CH}}-CH_2-O-(C_2H_4-O-)_{20}-CH_2-\overset{OH}{\underset{H}{\overset{|}{C}\cdot}} \longrightarrow$$

$$\underset{CN}{CH_2=CH} \longrightarrow \cdots-\left[\underset{CN}{\overset{}{CH}}-CH_2\right]_p-\overset{OH}{\underset{}{CH}}-CH_2-O-(C_2H_4-O-)_{20}-CH_2-\overset{OH}{\underset{}{CH}}-\left[CH_2-\underset{OH}{\overset{}{CH}}\right]_k-CH_2-\underset{CN}{\overset{}{CH}}-\cdots$$

This mechanism of initiation is confirmed by the fact that, when the PAN-PEO block copolymer is treated with diisocyanate in benzene in the presence of pyridine acting as catalyst, copolymers lose their solubility in DMF as a result of the formation of intermolecular chemical bonds[75].

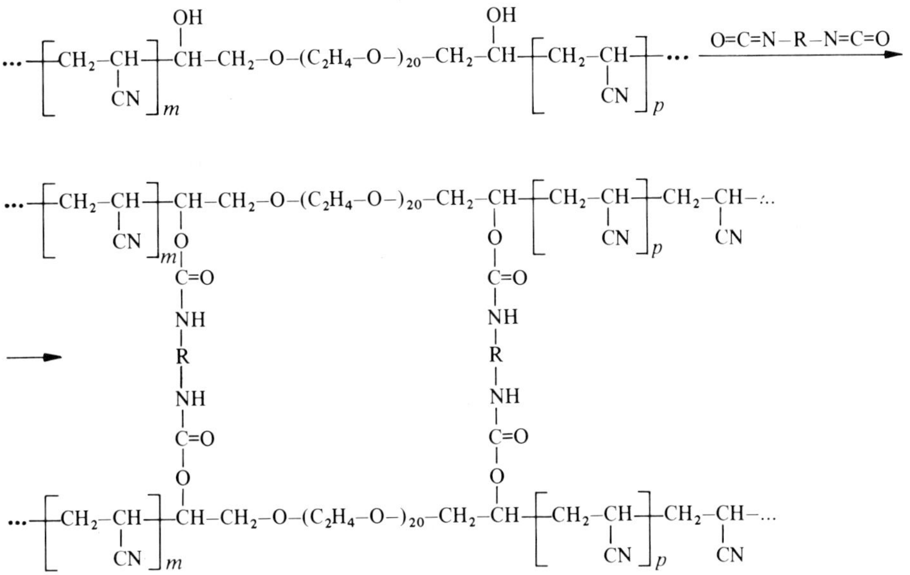

$$\cdots-\left[CH_2-\underset{CN}{\overset{}{CH}}\right]_m-\overset{OH}{\underset{}{CH}}-CH_2-O-(C_2H_4-O-)_{20}-CH_2-\overset{OH}{\underset{}{CH}}-\left[CH_2-\underset{CN}{\overset{}{CH}}\right]_p-\cdots \xrightarrow{O=C=N-R-N=C=O}$$

$$\cdots-\left[CH_2-\underset{CN}{\overset{}{CH}}\right]_m-\underset{\underset{\underset{\underset{\underset{O}{C=O}}{NH}}{R}}{C=O}}{\overset{}{O}}\text{-}CH-CH_2-O-(C_2H_4-O-)_{20}-CH_2-\underset{\underset{\underset{\underset{\underset{O}{C=O}}{NH}}{R}}{C=O}}{\overset{}{O}}\text{-}CH-\left[CH_2-\underset{CN}{\overset{}{CH}}\right]_p-CH_2-\underset{CN}{\overset{}{CH}}-\cdots$$

$$\longrightarrow \cdots-\left[CH_2-\underset{CN}{\overset{}{CH}}\right]_m-CH-CH_2-O-(C_2H_4-O-)_{20}-CH_2-CH-\left[CH_2-\underset{CN}{\overset{}{CH}}\right]_p-CH_2-\underset{CN}{\overset{}{CH}}-\cdots$$

The composition of block copolymers and, in particular, alternation of PAN and PEO blocks in the macromolecule of copolymers depends on the ratio between PEO and acrylonitrile in the reaction system. At low PEO concentrations in the reaction mixture a tri-block copolymer is probably formed with the following alternation of blocks:

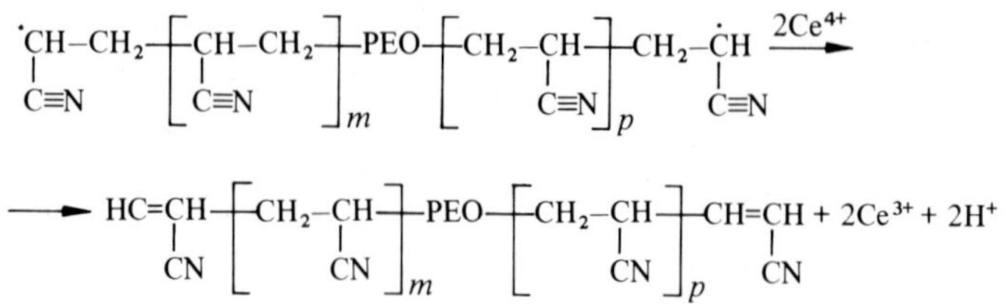

$$\overset{\bullet}{C}H-CH_2 {\Large[} \underset{C\equiv N}{\overset{|}{C}H}-CH_2 {\Large]}_m -PEO {\Large[} \underset{C\equiv N}{\overset{|}{C}H_2-CH} {\Large]}_p -\underset{C\equiv N}{\overset{|}{C}H_2-\overset{\bullet}{C}H} \xrightarrow{2Ce^{4+}}$$

$$\longrightarrow HC{=}CH {\Large[} \underset{CN}{\overset{|}{C}H_2-CH} {\Large]}_m -PEO {\Large[} \underset{CN}{\overset{|}{C}H_2-CH} {\Large]}_p -\underset{CN}{\overset{|}{C}H{=}CH} + 2Ce^{3+} + 2H^+$$

With the increase of PEO concentration in the reaction medium the number of active centres in PEO increases as a result of which the yield of block copolymer increases and, at the same time, a possibility arises for the chain to be disrupted, as shown in the scheme below, with the formation of a copolymer having the following structure:

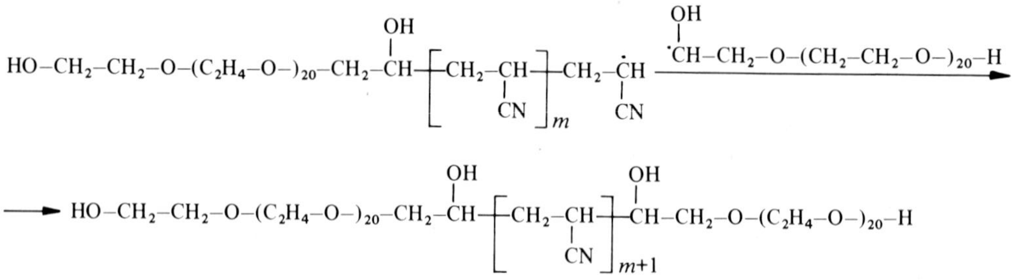

$$HO-CH_2-CH_2-O-(C_2H_4-O-)_{20}-CH_2-\underset{OH}{\overset{|}{C}H} {\Large[} \underset{CN}{\overset{|}{C}H_2-CH} {\Large]}_m -\underset{CN}{\overset{|}{C}H_2-\overset{\bullet}{C}H} \quad \underset{}{\overset{OH}{\overset{|}{C}H}}-CH_2-O-(CH_2-CH_2-O-)_{20}-H \longrightarrow$$

$$\longrightarrow HO-CH_2-CH_2-O-(C_2H_4-O-)_{20}-CH_2-\underset{}{\overset{OH}{\overset{|}{C}H}} {\Large[} \underset{CN}{\overset{|}{C}H_2-CH} {\Large]}_{m+1} \underset{}{\overset{OH}{\overset{|}{C}H}}-CH_2-O-(C_2H_4-O-)_{20}-H$$

Though the increase of the concentration of Ce^{4+} ions also increases the yield of copolymer, it facilitates at the same time the decrease of the length of polyacrylonitrile blocks due to cerium ions participating in the disruption of the growing chain.

The presence of flexible PEO and PPO blocks increases the viscosities of block copolymer solutions, this tendency is manifesting itself the stronger the greater is the PEO and PPO content in block copolymers.

Insertion of flexible blocks into the stiff polymer chain of PAN ensures the possibility of raising the resistance of modified PAN fibres to multiple deformations and, especially, increases significantly the abrasive resistance of fibres made of block copolymers.

5. References

[1] Gabrielyan, G. A., Rogovin, Z. A.: Khim. Volokna **1963** (5), 2; C. A. **60**, 5649 f
[2] Kol'k, A. R., Konkin, A. A., Rogovin, Z. A.: Izv. Akad. Nauk Est. SSR, Ser.-Fiz.-Mat. Tekh. Nauk, **1964** (3), 241; C. A. **62**, 7915 e
[3] Pen'kova, M. P., Konkin, A. A.: Khim. Volkna **1965** (3), 12; C. A. **63**, 8529 c
[4] USSR Auth. Gert. 306 139; C. A. **75**, P 141855 h
[5] Petrosyan, V. A., Gabrielyan, G. A., Rogovin, Z. A.: Khim. Volokna **1971** (1), 56; C. A. **74**, 113025 h
[6] Abramova, L. S., Gabrielyan, G. A., Rogovin, Z. A.: Vysokomol. Soedin., Ser. B **18**, 759 (1976)
[7] Zilberman, Ye. N.: Nitrile reactions. Moscow: Khimiya, 1972
[8] Iwekura, G., Nakabayashi, M., Lee, M.: Macromol. Chem. **78**, 157 (1964)
[9] Bose, A.: Ind. Eng. Chem. **32**, 16 (1940)
[10] Kol'k, A. R., Rogovin, Z. A., Konkin, A. A.: Khim. Volokna **1966** (6), 15; C. A. **66**, 38747 z
[11] US Pat. 2 842 525 (1958), B. F. Goodrich Co.
[12] Gabrielyan, G. A., Rogovin, Z. A.: Khim. Volokna **1963** (6), 2; C. A. **62**, 14871 c
[13] US Pat. 2 585 537 (1952), Coffman, D. (E. I. du Pont de Nemours and Co.)
[14] US Pat. 2 653 146 (1953), (E. I. du Pont de Nemours and Co.)
[15] Noro, K., Morimoto, G.: Kogyo Kagaki Zasshi **65**, 399 (1962); C. A. **57**, 10028 h
[16] Swakena, J., Tamikado, T., Fujimoto, Y.: Kobunsi Kagaki **15**, 469 (1958) C. A. **51**, 18692 f
[17] Gabrielyan, G. A., Stanchenko, G. I., Rogovin, Z. A.: Khim. Volokna **1965** (6), 13; C. A. **64**, 1865 b
[18] Schulz, R. C., Cherdon, H., Kern W.: Macromol. Chem. **24**, 141 (1957)
[19] Kol'k, A. R., Rogovin, Z. A., Konkin, A. A.: Khim. Volokna **1963** (4), 12, C. A. **59**, 11670 c
[20] Brandrup, F.: Faserforsch. Textiltech. **12**, 133, 208 (1961)
[21] Koral'nik, N. G.: Sb. Nauch. – issled. Robot, Khim i Khim. Tekhnol. Vysokomolekul. Soedin., Tashkentsk. Tekstiln. inst., **1964**, p. 46; C. A. **63**, 17882 d
[22] Kol'k, A. R., Konking, A. A., Rogovin, Z. A.: Izv. Akad. Nauk Est. SSR, Fiz.-Mat. Tekh. Nauk **1964** (3) 246; C. A. **62**, 16431 g
[23] Pen'kova, M. P., Konkin, A. A.; Khim. Volokna **1965** (3), 12; C. A. **63**, 8529
[24] USSR Auth. Cert. 166095; C. A. **62**, P6614 g
[25] Kudryavtsev, G. I., Vasilyevy, Ye. A., Zharkova, M. A.: Zh. Prikl. Khim. (Leningrad) **32**, 594 (1958); C. A. **52**, 21213 e
[26] Price, C. C., Zomlefer, Y.: J. Am. Chem. Soc. **72**, 14 (1950)
[27] Ham, D. (ed.): Copolymerization (Transl. form Engl). Moscow: Khimiya, 1971, pp. 360–374
[28] Vialle, J., Gulot, J., Gugot, A.: J. Macromol. Sci., Chem. **5**, 1031 (1971)
[29] US Pat. 3 296 228 (1967) (E. I. du Pont de Nemours and Co.)
[30] Petrosyan, V. A., Gabrielyan, G. A. Rogovin, Z. A.: Khim. Volokna, 1972 (1), 68; C. A. **76**, 142210 m
[31] USSR Auth. Cert, 450 002 (1974); C. A. **82**, 157711 y
[32] USSR Auth. Cert., 359 311; C. A. **78**, 137809 a
[33] USSR Auth. Cert., 456 057 (1974); C. A. **83**, 19744 y
[34] Vyong-Kong, Li, Gabrielyan, G. A., Pen'kova, M. P., Rogovin, Z. A.: Khim. Volokna **1973** (4), 20; C. A. **82**, 44776 b
[35] Logan, H., Nichols, R.: J. Res. Nat. Bur Stand. **41**, 521 (1948)
[36] Drabkin, I. A.: Dokl. Akad. Nauk SSSR **154**, 197 (1964)
[37] US Pat. 241 1599 (1940); C. A. **41**, 6080 e. (Standard Oil Co.)
[38] Ger. (East) Pat. 50946 (1966); C. A. **66**, 66263 g
[39] US Pat. 2 786 043 (1955) (American Cyanamid Co.); C. A. **51**, 11758 i
[40] USSR Auth. Cert. 351 860; C. A. **78**, 177809 a
[41] USSR Auth. Cert. 398 557 (1973); C. A. **81**, 121381 f
[42] Zinina, A. P., Pen'kova, M. P., Gabrielyan, G. A., Rogovin, Z. A.: Vysokomol. Soedin., Ser. B, **15** (7), 496 (1973); C. A. **80**, 37504 v

[43] Zhovner, N., Zakharov, N. P., Orekhov, S. V.: Izv. Vyssh. Uchebn. Zaved., Khim. Khim. Tekhnol. **13**, 152 (1970); C. A. **73**, 131779r
[44] Ham, G.: J. Polym. Sci. **54**, 411 (1961)
[45] Zharkova, M. A., Kudryavtsev, V. I.: Khim. Volokna **1969** (2), 49; C. A. **71**, 14067e
[46] Kamalov, S., Gladkikh, A. F., Frenkel, S. Ya.: Khim. Volokna **1967** (3) 21., C. A. **68**, 3786a
[47] Gabrielyan, G. A., Pen'kova, M. P., Zinina, A. P., Chernukhina, A. I.: Prepr. Int. Symp. Chem. Fibres **3**, 82 (1974)
[48] Kabanov, V. A., Kargina, O. V., Petrovskaya, V. A.: Vysokomol. Soedin., Ser. A **13**, 348 (1971); C. A. **74**, 112456n
[49] Kabanov, V. A., Topchiev, D. A.: Polymerization of ionized monomers Moscow: Khimiya 1975, p. 51
[50] Konnova, N. F., Gabrielyan, G. A., Konkin, A. A., Rogovin, Z. A.: Vysokomol. Soedin 7, 756 (1965); C. A. **63**, 5764h
[51] Konnova, N. F., Gabrielyan, G. A., Konkin, A. A., Rogovin, Z. A.: Vysokomol. Soedin. 8, 422 (1966); C. A. **64**, 19789a
[52] Konnova, N. F.: Auth. Abstr. of Cand. Thesis, Moscow 1967
[53] Pastushak, N. O., Dombrovsky, A. K.: Zh. Org. Khim. **2**, 324 (1965); C. A. **62**, 16099h
[54] Gabrielyan, G. A. Rogovin, Z. A.: Vysokomol. Soedin. **6**, 769 (1964); C. A. **61**, 5801h
[55] Chukhadzhian, G. A., Kalaidzhian, A. Ye., Petrosyan, V. A.: Vysokomol. Soedin., Ser. A **12**, 171 (1970); C. A. **72**, 2303m
[56] Levites, L. M., Gabrielyan, G. A., Kudryavtsev, G. I., Rogovin, Z. A.: Vysokomol. Soedin., Ser. B **12**, 309 (1970); C. A. **73**, 35912h
[57] Shalabi, S., Gabrielyan, G. A., Konkin, A. A.: Vysokomol. Soedin., Ser. B, **12**, 6 (1970); C. A. **74**, 54940c
[58] USSR Auth. Cert. **186**, 622 (1966); C. A. **66**, P 66612v
[59] Shalabi, S., Gabrielyan, G. A., Konnova, N. F., Konkin, A. A., Smutkina, Z. S.: Vysokomol. Soedin, Ser. A **16** (8) 1904 (1974); C. A. **82**, 59604w
[60] Levites, L. M., Gabrielyan, G. A., Kudriavcev, G. I., Rogovin, Z. A.: Faserforsch. Textiltech. **25**, 153 (1974)
[61] Madicke, A., Krahnstover, M., Sauerwald, F.: Faserforsch. Textiltech. **19**, 401 (1968)
[62] Rogovin, Z. A.: New. cellulose materials. Moscow: Znaniye 1967, pp. 3–60
[63] Lits, N. P., Druzhinina T. V.: Khim. Volokna **1975** (2), 15; C. A. **82**, 87471u
[64] Wolf, L. A., Meos, A. I.: Special purpose fibres. Moscow: Khimiya, 1971, pp. 59–136
[65] USSR Auth. Cert. 166 095 (1964); C. A. **62**, 6644g
[66] Pen'kova, M. P., Konkin, A. A., in: Carbochain fibres, Moscow: Knimiya, 1966
[67] Pen'kova, M. P., Skorobogatova, L., Konkin, A. A.: Khim Volokna **1966** (2), 18
[68] El Garf, S., Konkin, A. A., Rogovin, Z. A.: Vysokomol. Soedin. 8, 72 (1966)
[69] Shalabi, S., Gabrielyan, G. A., Konkin, A. A.: Khim. Volokna **1971** (6) 43; C. A. **76**, 113 802a
[70] Chernukhina, A. I., Gabrielyan, G. A., Rogovin, Z. A.: Vysokomol. Soedin., Ser. B **16** (11) 817 (1974); C. A. **82**, 125 768h
[71] Chernukhina, A. I., Gabrielyan, G. A., Rogovin, Z. A.: Khim Volokna **1974** (2), 16. C. A. **83**, 80956t
[72] Grassie, N., Neill, J.: J. Chem. Soc. **1966**, 3929
[73] Novitskaya, M. A., Konkin, A. A.: Vysokomol. Soedin. **7**, 1719 (1965); C. A. **64**, 3710g
[74] Novitskaya, M. N., Konkin, A. A., in: Carbochain fibres, Moscow: Khimiya, 1966, p. 246; C. A. **67**, 100890s
[75] Novitskaya, M. N., Konkin, A. A.: Trans. Sib. Technol. Inst., Krasnoyarsk **1966** (2), 37; C. A. **68**, 88053g
[76] USSR Auth. Cert., 431 182; C. A. **83**, 80956t

Received February 2, 1977
W. Kern (editor)

Synthetic Polyelectrolytes as Models of Nucleic Acids and Esterases

by Tsuneo Okubo and Norio Ise

Department of Polymer Chemistry, Kyoto University, Kyoto, Japan

Table of Contents

1. Introduction 136
2. Nucleic Acid Models 136
2.1. Anionic Polyelectrolytes Containing Nucleic Acid Bases 136
2.2. Cationic Polyelectrolytes Containing Nucleic Acid Bases 139
2.3. Neutral Water-Soluble Polymers Containing Nucleic Acid Bases 143
2.4. Ionic Surfactants Containing Nucleic Acid Bases 146
2.5. Affinity Chromatography by Resin-Type Models 148
2.6. Template-Directed Polymerization of Nucleotides 151
3. Esterase Models 154
3.1. Polyelectrolyte-Catalyzing Acid Hydrolysis 155
3.2. Polyelectrolyte-Catalyzing Alkaline Hydrolysis 157
3.3. Polyelectrolytes Containing Nucleophile Groups 162
3.4. Esterolytic Resin-Type Polyelectrolytes 168
3.5. Basic Aspects of Polyelectrolytes Retaining Esterase Activities 172
4. Conclusion 176
5. References 178

1. Introduction

Polyelectrolytes are polymers containing dissociable groups. The huge electrostatic potential field displayed by macroions is a source of various characteristic properties of polyelectrolytes in solutions. So far a large number of investigations have been carried out on the characteristic properties of polyelectrolytes both in solution and solid states. Thermodynamic properties such as activity coefficient, osmotic coefficient, heat of dilution, partial molal volume, etc. as well as hydrodynamic properties (viscosity, transference number) in aqueous media have been most intensively investigated up to present. Furthermore, keen attention has been paid also to syntheses of polyelectrolytes as models of naturally occurring substances such as proteins, nucleic acids, enzymes, carbohydrates, etc. On the basis of their fundamental chemical structures, a variety of biologically important polymers are regarded as polyelectrolytes. Thus, it has been considered to be most useful and interesting to clarify the complicated function of biological substances by making use of synthetic polyelectrolytes of more simplified chemical structures. The function of enzymes may be characterized by (1) high efficiency, (2) high specificity, and (3) regulative mechanism. Although all of these should be targets for model polyelectrolytes, the present status of the research is far from satisfactory. In the following chapters, polyelectrolytes as models of nucleic acids and esterases hitherto investigated will be described. As would be clear, the specificity, if we are allowed to use this expression, has been visualized by the model compounds to "some extent". The rather high efficiency of the model polyelectrolytes has been achieved, but will not be the main subject of the present article. The readers are referred to Refs.[1-3] on this topics. With respect to the model compounds of the nucleic acid (Chapter 2), Ref.[4] is a review to be referred to and earlier work on the esterase models has been reviewed in Refs.[5-8].

2. Nucleic Acid Models

Nucleic acids, like enzyme proteins, are macromolecules having dissociable charges, *i.e.*, polyelectrolytes. The molecular weight of nucleic acids is very high, ranging from 10^6 to 10^9 or more. Chemically, they are composed of phosphate groups, heterocyclic nucleic bases (adenine, cytosine, guanine, and thymine), and pentoses. Many researchers have synthesized a variety of model compounds, which contain all or part of these ionic groups, bases and pentose groups.

2.1. Anionic Polyelectrolytes Containing Nucleic Acid Bases

Kinoshita, Imoto et al.[9-12] polymerized 3-(9-adenyl)-2-hydroxypropyldihydrogenphosphate, *1* (oligo-9-AP), 3-(3-adenyl)-2-hydroxypropyldihydrogenphosphate, *2* (oligo-3-AP), 3-(9-hypoxanthyl)-2-hydroxypropyldihydrogenphosphate, *3* (oligo-9-HP), and 3-(3-hypoxanthyl)-2-hydroxydihydrogenphosphate, *4* (oligo-3-HP) by condensation with N,N'-dicyclohexylcarbodiimide. However, the molecular weight was very low ($\simeq 10^3$). The absorption spectra showed an apparent hypochromicity of 3% for oligo-9-AP + RNA (DNA) mixtures, which was calculated according to the following equation[13],

$$\text{Hypochromicity}(\%) = 100 \times \left(1 - \frac{I_{a+b}}{mI_a + nI_b}\right) \qquad (1)$$

where m and n are the volume fractions of polymer solutions a and b, and I_a, I_b and I_{a+b} are the absorbances of the solutions of the component polymers and the mixture solutions, respectively.

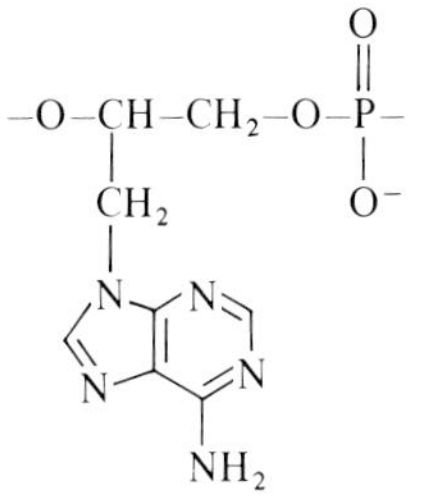

1, oligo–9–AP

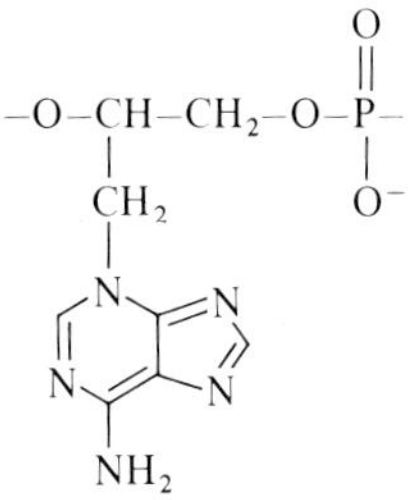

2, oligo–3–AP

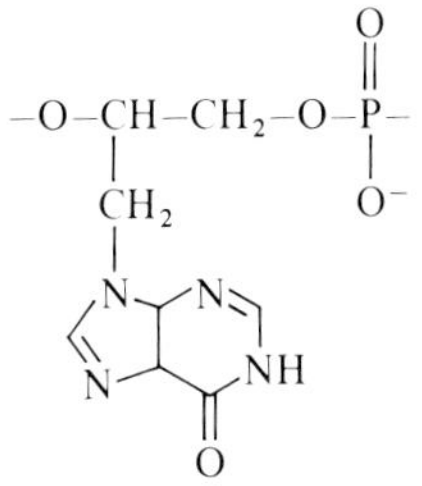

3, oligo–9–HP

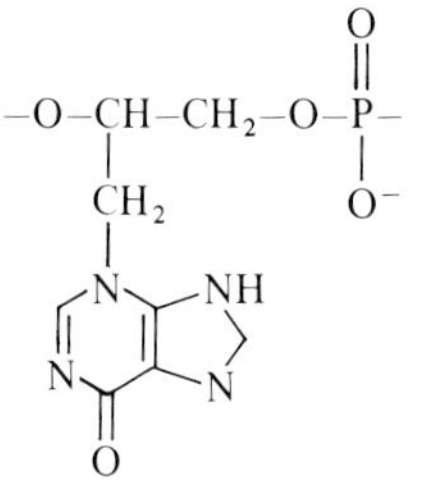

4, oligo–3–HP

Kinoshita, Imoto et al.[11, 14] synthesized other anionic models, 5 (APVP), CPVP, UPVP, TPVA, HPVA, THPVA, and 6 (AMPPVA), by the polymer reaction of N-coupled(2-dihydrogenphosphate)-ethylderivatives of nucleic acid bases (or adenosine-5'-phosphate, AMP) with polyvinylalcohol. A, C, U, T, H, and TH denote adenine, cytosine, uracil, thymin, hypoxanthine, and theophylline, respectively. The authors reported the apparent hypochromities of 3 to 16% for many kinds of mixtures of the models and DNA or RNA, as compiled in Table 1. However, for the mixtures APVA + RNA, HPVA + RNA, HPVA + DNA, THPVA + RNA, CPVA + DNA and CPVA + RNA, no hypochromicity was detected.

Polyvinylalcohol containing two bases, i.e., ATPVA and AUPVA, were prepared by Kinoshita, Imoto et al.[12], in which the absorbance increased by 7~8% with temperature rising from 20° to 50° as is shown in Fig. 1.

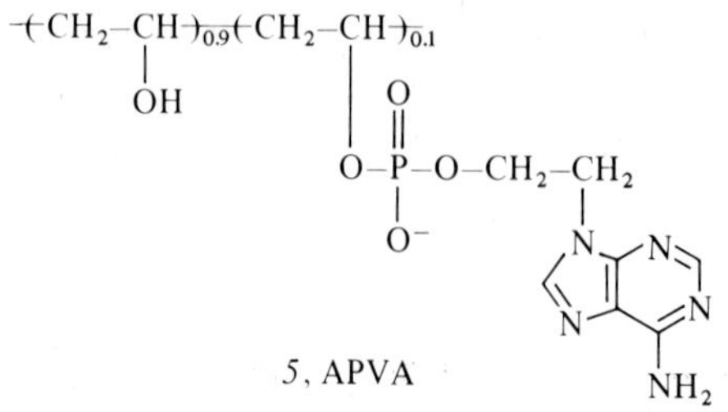

$\pm CH_2-CH\xrightarrow{}_{0.9}(CH_2-CH)_{0.1}$
  | |
  OH   O
                       ‖
                       O–P–O–CH$_2$–CH$_2$
                       |
                       O$^-$

5, APVA

—CH$_2$–CH–CH$_2$–CH—
    | |
    OH   O
             |
             O=P–O$^-$
             |
             O–CH—[ribose]—adenine
                HO OH

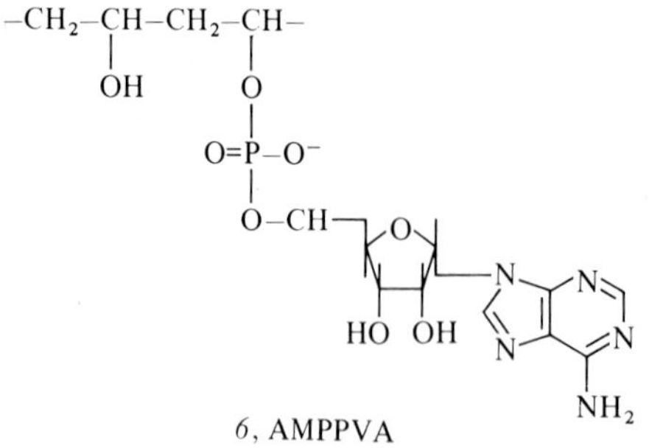

6, AMPPVA

Table 1. Apparent hypochromicities of anionic models of nucleic acid

Mixtures	Solvent	λ_{max} (nm)	Hypochro- micity (%)	Ref.
Oligo-9-AP + RNA	0.1 MNa$_2$HPO$_4$ + H$_2$O	260	3	10)
Oligo-9-AP + DNA	0.1 MNa$_2$HPO$_4$ + H$_2$O	260	3	10)
Oligo-3-AP + RNA	0.1 MNa$_2$HPO$_4$ + H$_2$O	265	0	10)
Oligo-3-AP + DNA	H$_2$O	265	0	10)
Oligo-9-AP + RNA	0.1 MNa$_2$HPO$_4$ + H$_2$O	260	0	10)
APVA + RNA	0.1 MNa$_2$HPO$_4$ + H$_2$O	260	0	11)
APVA + DNA	0.1 MNa$_2$HPO$_4$ + H$_2$O	260	6.5	11)
TPVA + DNA	0.1 MNa$_2$HPO$_4$ + H$_2$O	266	16	11)
TPVA + RNA	0.1 MNa$_2$HPO$_4$ + H$_2$O	266	12	11)
APVA + TPVA	0.1 MNa$_2$HPO$_4$ + H$_2$O	265	3	11)
HPVA + RNA	0.1 MNa$_2$HPO$_4$ + H$_2$O	259	0	11)
HPVA + DNA	H$_2$O	259	0	11)
THPVA + RNA	H$_2$O	268	0	11)
CPVA + DNA	H$_2$O	267	0	14)
CPVA + DNA	0.1 MNa$_2$HPO$_4$ + H$_2$O	267	0	14)
CPVA + RNA	0.1 MNa$_2$HPO$_4$ + H$_2$O	267	4	14)
CPVA* + RNA	0.1 MNa$_2$HPO$_4$ + H$_2$O	267	0	14)
UPVA + DNA	0.1 MNa$_2$HPO$_4$ + H$_2$O	261	3	14)
UPVA + RNA	0.1 MNa$_2$HPO$_4$ + H$_2$O	261	3	14)
UPVA* + RNA	0.1 MNa$_2$HPO$_4$ + H$_2$O	261	0	14)
AMPPVA + RNA	H$_2$O	260	3	14)
AMPPVA + RNA	0.1 MNa$_2$HPO$_4$ + H$_2$O	260	3	14)
AMPPVA + DNA	0.1 MNa$_2$HPO$_4$ + H$_2$O	260	4	14)
AMPPVA* + RNA	0.1 MNa$_2$HPO$_4$ + H$_2$O	260	0	14)

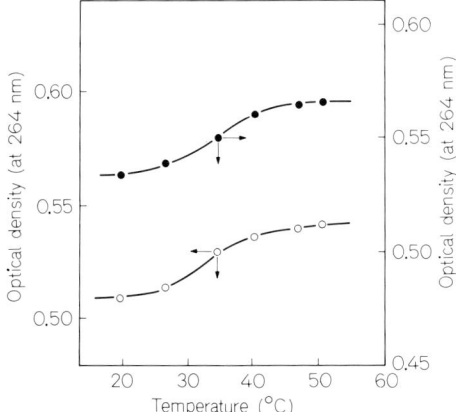

Fig. 1. Temperature dependence of optical density. ●: ATPVA (at 274 nm), ○: AUPVA (at 264 nm) (Ref.[12])

For the anionic models, the detailed studies on their interactions with poly- and mononucleotides have not yet been carried out. However, we may say that the interactions of bases between the models and polynucleotides would be faint because of the strong electrostatic repulsive forces between the macroanions.

2.2. Cationic Polyelectrolytes Containing Nucleic Acid Bases

Nucleic acids are anionic under the neutral conditions. Thus, the syntheses of model compounds of the opposite charge are interesting for the discussion of electrostatic contributions in specific interactions of nucleic acids. We have tried to synthesize cationic models by the Menschutkin reaction of poly-4-vinylpyridine with 9-(2-chloroethyl)adenine, 1-(2-chloroethyl)thymine, and 7-(2-chloroethyl)theophylline[15, 16]. The obtained polymers are poly 1-[2-(adenin-9-yl)ethyl]-4-pyridinioethylene chloride 7 (APVP), poly 1-[2-(thymin-1-yl)ethyl]-4-pyridinioethylene chloride 8 (TPVP), and poly 1-[2-(theophyllin-7-yl)ethyl]-4-pyridiniothylene chloride 9 (THPVP), respectively.

7, APVP   8, TPVP   9, THPVP

An aqueous solution of APVP was highly viscous, and the reduced viscosity was rather insensitive to polymer concentration, unlike that of usual synthetic linear polyelectrolytes[17, 18]. This suggests that the model compounds were rather stiff in solution, which would be due to the bulkiness and the hydrophobicity of side groups. The stiffness of APVP was also demonstrated by similar insensitivity of the viscosity toward temperature.

The changes in UV-absorbance with a mixed ratio of mixtures of the present model compounds are shown in Fig. 2.

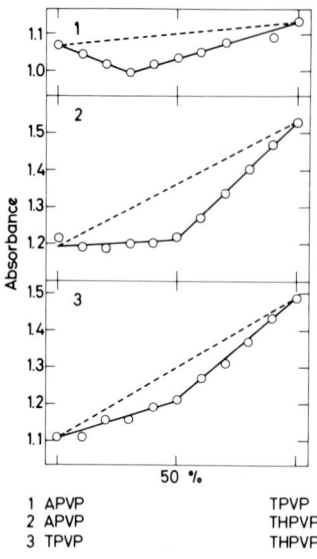

Fig. 2. Absorbance of APVP + TPVP, APVP + THPVP, and TPVP + THPVP mixtures in Walpole acetate buffer (pH = 4.5). Absorbances at 255 nm (1 and 2) and 263 nm (3) were observed at 25 °C (Ref.[16])

It can be seen from the figure that the electrostatic repulsive forces between the macrocations are overwhelmed, probably by hydrophobic attractive forces between their hydrophobic side groups. It should be noted that the complimentary base-base pairing is unimportant in the present case. If this is not the case, the mixtures of APVP and TPVP should show the largest hypochromicity. This, however, is not the case. The importance of the hydrophobic interactions between nucleic acid bases has been proposed by Ts'o et al.[19] from thermodynamic parameters of various nucleic acid bases or nucleosides in aqueous media.

Recently, similar model compounds were synthesized by Shimidzu et al.[20]. Furthermore, Shimidzu et al.[20] aimed at synthesizing other cationic polymers: poly 1-[3-(adenin-9-yl)propyl]-4-pyridinioethylene chloride, APVP*, poly 1-[2-(adenin-9-yl)ethyl]-iminoethylene halide, 10 (APEI), poly[1-(2-thymin-1-yl)-iminoethylene halide], 11 (TPEI), poly[1-(uracil-5-yl)-iminoethylene halide], 12 (UPEI), and poly[1-(uridin-5'-yl)-iminoethylene halide], 13 (UPEI*). They were obtained by quaternization of poly-4-vinylpyridine or polyethylenimine with chloroethylated, chloropropylated, or chlorated nucleic acid bases.

The apparent hypochromicities of cationic analogs with polynucleotides are compiled in Table 2. As is clear from the table, some of the cationic models interact strongly with polynucleotides. The hypochromicities of the polymers with polynucleotides are similar to that reported for some neutral model compounds[21-23], or fairly higher than those of most neutral and anionic models hitherto synthesized (see Sections 2.1. and 2.3.). The large hypochromicity values of the cationic models

Table 2. Apparent hypochromicities of cationic models of nucleic acid

Mixtures	Solvent	λ_{max} (nm)	Hypochromicity (%)	Ref.
APVP + TPVP	Walpole acetate buffer	255	8	16)
APVP + THPVP	Walpole acetate buffer	255	11	16)
TPVP + THPVP	Walpole acetate buffer	255	6	16)
APVP + Poly A	KCl(0.1 M) + HCl(0.1 M) + H_2O	255	17	16)
APVP + Poly A	KCl(0.2 M) + H_2O	255	21	16)
APVP + Poly U	NaBr(5 M) + H_2O	263	8	16)
APVP + DNA	NaBr(5 M) + H_2O	255	15	16)
APVP + AMP	H_2O	259	12	20)
APVP + GMP	H_2O	255	8	20)
APVP + CMP	H_2O	265	7	20)
APVP + UMP	H_2O	260	5	20)
APVP + TMP	H_2O	263	6	20)
TPVP + AMP	H_2O	262	10	20)
TPVP + GMP	H_2O	258	9	20)
TPVP + CMP	H_2O	268	4	20)
TPVP + UMP	H_2O	263	5	20)
TPVP + TMP	H_2O	266	3	20)
APVP + NAD	Walpole acetate buffer	255	1	16)
TPVP + NAD	Walpole acetate buffer	265	8	16)
THPVP + NAD	Walpole acetate buffer	265	2	16)
APVP + adenine	H_2O	260	2	20)
APVP + thymin	Walpole acetate buffer	255	2	16)
APVP + uracil	H_2O	261	2	20)
TPVP + adenine	Walpole acetate buffer	255	3	16)
TPVP + adenine	H_2O	260	6	20)
TPVP + guanine	H_2O	260	6	20)
TPVP + uracil	H_2O	264	2	20)
APEI + AMP	H_2O	260	7	20)
APEI + GMP	H_2O	258	8	20)
APEI + CMP	H_2O	267	4	20)
APEI + UMP	H_2O	262	2	20)
APEI + TMP	H_2O	265	3	20)
UPEI + AMP	H_2O	260	1	20)
UPEI + UMP	H_2O	260	1	20)
UPEI* + AMP	H_2O	260	5	20)
UPEI* + GMP	H_2O	260	4	20)
UPEI* + UMP	H_2O	260	4	20)

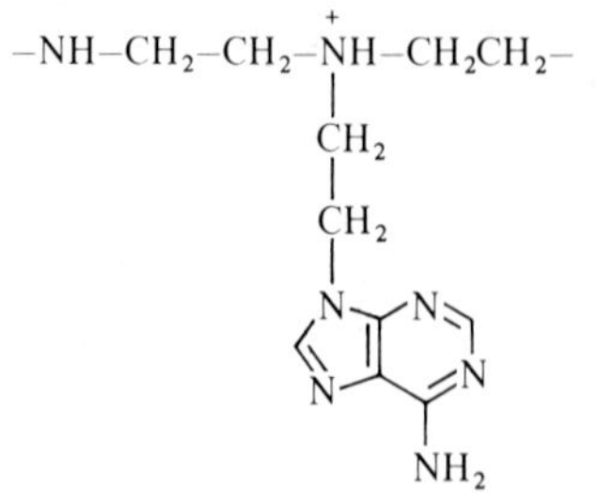

10, APEI

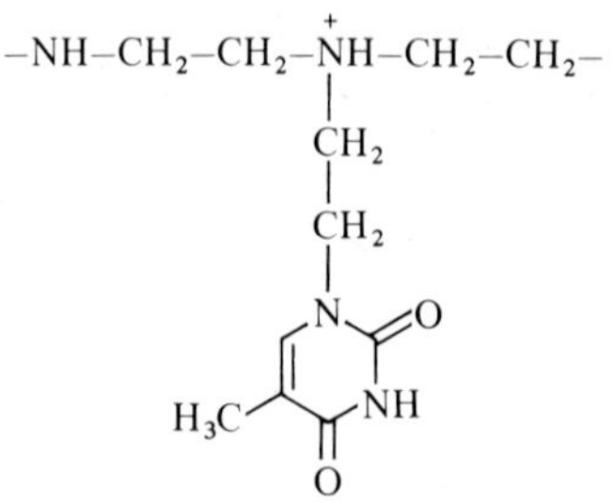

11, TPEI

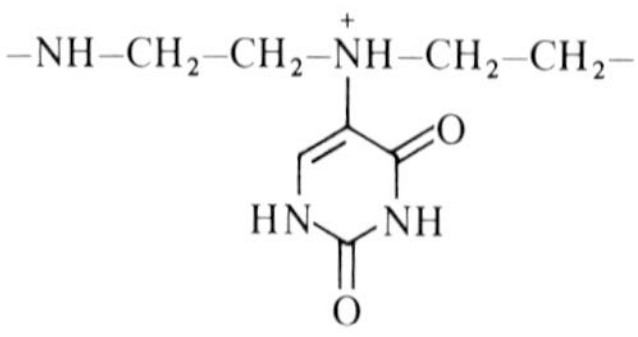

12, UPEI

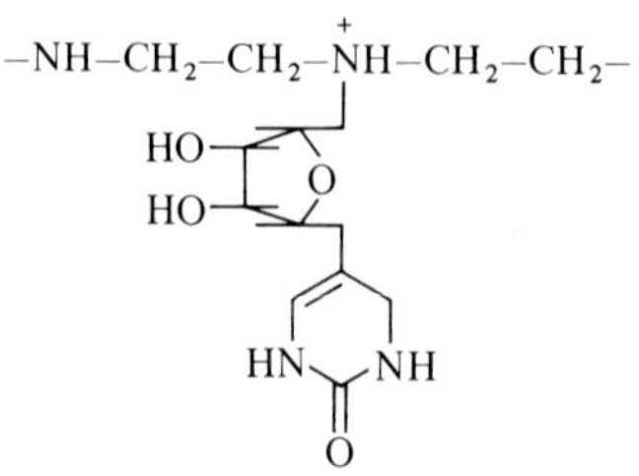

13, UPEI*

may be due to strong electrostatic attractive forces between macrocations and macroanions (polynucleotide). The interactions of APVP or TPVP with mononucleotides were weaker than those with polynucleotides. The apparent hypochromicities with nucleosides or nucleic acid bases were still smaller. Furthermore, an apparent hypochromicity between APVP and Poly A sharply decreased with increasing ionic strength, and the interactions of the quaternized pyridine with mononucleotides were quite feasible. All these results support the significance of electrostatic interactions.

A hypochromicity was observed between THPVP and APVP (or TPVP). Since theophylline is not a nucleic acid base and does not form hydrogen-bonding, these observations indicate that stacking-type hydrophobic forces are important.

It should be noted here that the hypochromicities of the models with purine bases or purine nucleotides seem to be larger than those with pyrimidine bases or nucleotides in Table 2. This feature suggests that the hydrophobic interactions are predominant and the hydrogen-bonding is of secondary importance. The absorption-temperature profiles of APVP, TPVP, and THPVP are shown in Fig. 3. It was found that the absorbance decreased linearly with increasing temperature, between 10 and 80 °C, rather sharply for APVP and slightly for TPVP and THPVP. These temperature dependences are the opposite of those reported for ordinary polynucleotides[24]. The present results can be explained by the characteristic temperature dependence of the hydrophobic interactions; their strength increases with increasing temperature, as proposed by Scheraga et al.[25]. The absorbances of APVP – THPVP mixtures also decreased linearly with elevating temperature, reflecting that the stacking of nucleic acid bases is not due to hydrogen-bonding but to hydrophobic interactions.

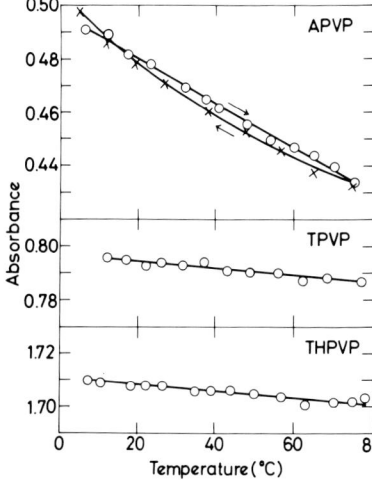

Fig. 3. Temperature dependence of the absorbance of APVP, TPVP, and THPVP (Ref.[16])

2.3. Neutral Water-Soluble Polymers Containing Nucleic Acid Bases

One of the simplest synthetic models of nucleic acid may be poly-9-vinyladenine, *14* (PA) or poly-1-vinyluracil, *15* (PU). These models were synthesized independently by Pitha et al.[23, 26–30], Takemoto et al.[21, 31–33], and Kaye et al.[22, 34–39]. The hypochromicity of PA and PU with polynucleotides is listed in Table 3.

Kaye and Chang[38] found that the highly syndiotactic PU had a larger hypochromism than the less syndiotactic ones in aqueous media. The authors found interesting results on the thermal melting of PU-PA complexes (see Fig. 4). The melting

14, PA   *15*, PU

curve of PA-PU complexes showed an decreasing dependence of absorbance on temperature. The reason may be the base-base stacking stimulated by hydrophobic interactions between bases.

Kaye and Chou[39] also studied the effect of base stacking on the conformation of PA using osmometry, intrinsic viscosity, and light-scattering. The ideal behavior (under the θ conditions) of PA existed at neutral pH (= 7.4) and at 26 and 40 °C from the osmotic measurements.

Intrinsic viscosity measurements revealed a conformational transition upon heating from 26 to 40 °C, while the UV absorbance of the solution was insensitive to the change. The entropy parameters for PA were also discussed in light of the Flory-Krigbaum correlation between the second virial coefficient and theta temper-

Table 3. Apparent hypochromicities of neutral models of nucleic acid

Mixtures	Solvent	λ_{max} (nm)	Hypochromicity (%)	Ref.
PA + RNA	NaCl(1 M) + H_2O	258	13	21)
PA + RNA	H_2O	258	13	21)
PA + Poly U	Sodium cacodylate buffer	260	28	27)
PA + PU	trimethylphosphate + H_2O	258	5	21)
PAU + DNA	H_2O	260	20	40)
PPAOU + Poly A	NaCl(0.15 M) + trisodium citrate (0.015 M) + H_2O	260	5	42)
PPAOU + DNA	NaCl(0.15 M) + trisodium citrate (0.015 M) + H_2O	260	5	42)
CMCT + Poly A	H_2O	266	5	43)
CMCT + Poly A	Glycine buffer (pH = 6.3)	266	5	43)
CMCT + Poly U	Glycine buffer (pH = 6.3)	266	0	43)
CMCT + DNA	Glycine buffer (pH = 6.3)	266	0	43)
P·TA + RNA	H_2O	260	0	21)
PMAOA + PMAOT	Trimethylphosphate + H_2O	260	3	21)
PMAOA + PMAOU	Trimethylphosphate + H_2O	260	3	21)
PMAOT + PMAOU	Trimethylphosphate + H_2O	260	0	21)
PU·MA + PA	H_2O	257	0	21)
PA·A + PT·A	H_2O	260	0	21)

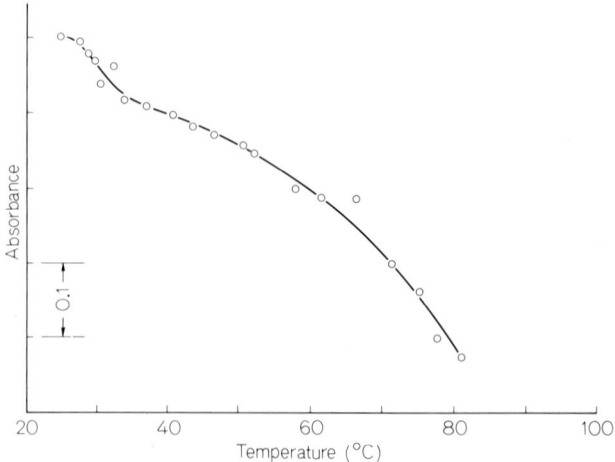

Fig. 4. UV-absorbance melting profile of 2 PU + PA complex in 0.01 M sodium cacodylate. pH = 7.0, in 0.01 M NaCl (Ref.[38])

ature. The transition was accompanied by an inversion in the sign of the entropy parameter from negative to positive upon rising temperature. From the above results Kaye et al. suggested that both nearest-neighbour and long-range intramolecular arrangements of base-stacking occur in PA.

Jones and co-workers synthesized copolymers of acrylamide containing nucleoside as a neutral model, i.e. poly(5'-O-acrylyluridine-co-acrylamide), 16 (PAU), poly(5'-O-acrylylthymidine-co-acrylamide), PAT[40], poly(5'-O-acryloyluridine-co-acrylamide), 17 (PAOU)[41], poly[5'-O-(3-vinylacryloyl)uridine] (PAOU*), poly[5'-O-(β-vinylacryloyl)uridine-co-acrylamide], 18 (PPAOU)[42], and poly(3'-O-carboxymethylthymidine), 19 (CMCT)[43]. Hypochromicities were observed for the mixtures of PAU + DNA, PPAOU + Poly A, PPAOU + DNA and CMCT + Poly A

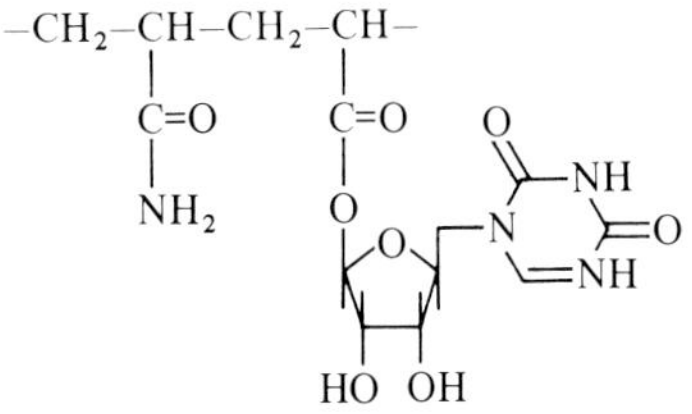

16, PAU

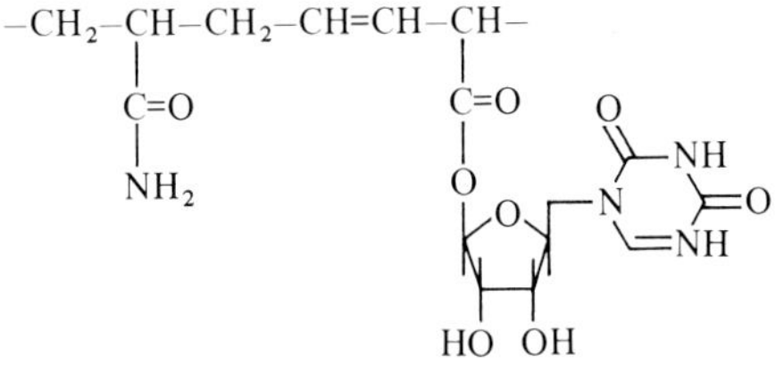

17, PAOU

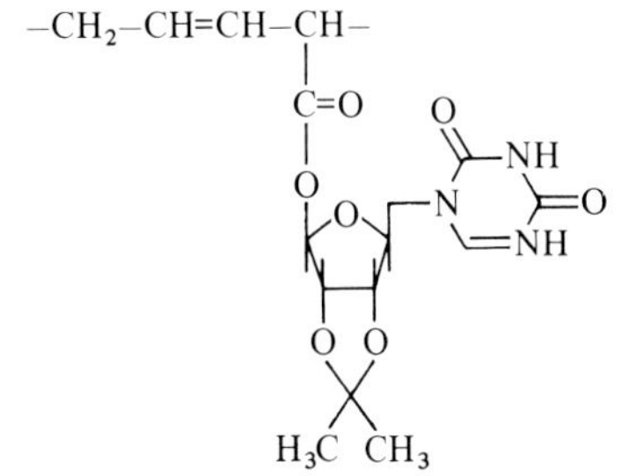

18, PPAOU

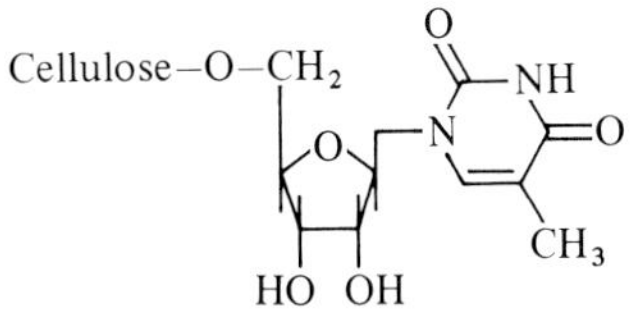

19, CMCT

(see Table 3). From these results, it is not clear whether the interactions are due to the hydrogen-bonding or to the hydrophobic base-stacking.

Takemoto et al.[21] synthesized similar types of models, i. e. poly(β-methacryloyloxyethyladenine) 20 (PMAOA), poly(β-methacryloyloxyethyluracil) (PMAOU), and poly(β-methacryloyloxyethylthymine) (PMAOT), by free-radical polymerizations of β-methacryloxyethyl compounds of the corresponding base.

Copolymers of 9-vinyladenine and acrylamide, 21 (PA·A), 1-vinylthymine and acrylamide, 22 (PT·A) and 1-vinyluracil and maleic anhydride, 23 (PU·MA) were also synthesized[21]. The hypochromicities with RNA were measured and compiled in Table 3.

Peptide-type models were synthesized by Jones et al.[44], and by Takemoto et al.[45]. Takemoto and co-workers[46] prepared polymers of 6-methylamino-, 6-di-n-propylamino-, and 6-di-n-butylamino-9-vinylpurines. They reported that these models showed no hypochromic effect on RNA in aqueous solution.

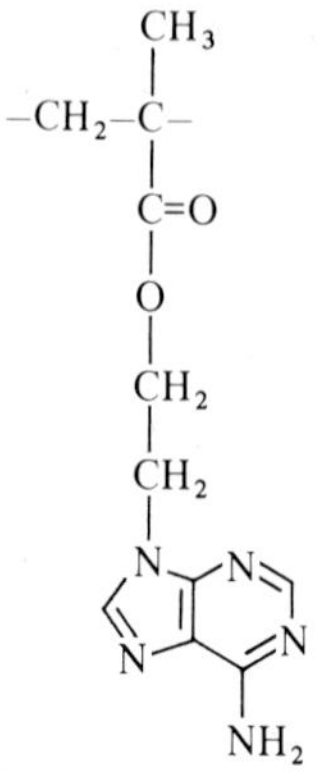

20, PMAOA

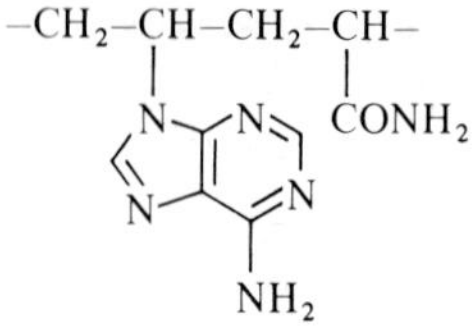

21, PA·A

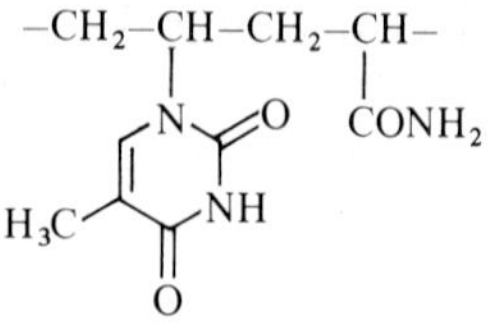

22, PT·A

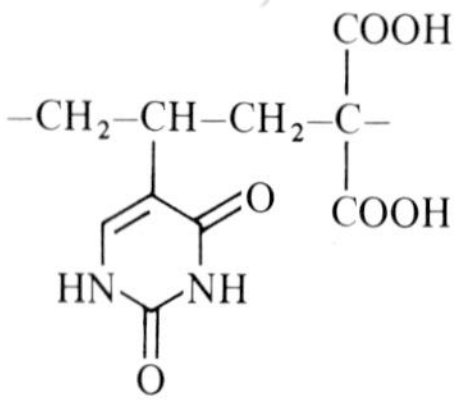

23, PU·MA

2.4. Ionic Surfactants Containing Nucleic Acid Bases

From spectroscopic studies on the interactions of linear polymer-type models with polynucleotides and with related compounds, it was found that the hydrogen-bond formation between complementary bases was not a main factor for the hypochromic effects, but the hydrophobic interactions were suggested to be more important. If ionic surfactant models are used, contributions of the hydrophobic, electrostatic and hydrogen-bond interactions to the base-base pairing and/or stacking of nucleic acid bases can be discussed in more detail. Furthermore, it is interesting to examine the correlation between base-base interactions and conformation, because the models form micelles above a critical micell concentration. We synthesized five kinds of cationic surfactants: β-(adenin-9-yl)ethyldodecyldimethylammonium chloride, 24 (A 12), β-(thymin-1-yl)ethyldodecyldimethylammonium chloride), 25 (T 12), β-(theophyllin-7-yl)ethyldodecylammonium chloride, 26 (TH 12), β-(adenin-9-yl)-ethyloctadecyldimethylammonium chloride (A 18), and β-(thymin-1-yl)ethyloctadecyldimethylammonium chloride (T 18)[47]. These model compounds were prepared by the Menschutkin reaction of dodecyldimethyl amine or octadecyldimethylamine with chloroethylated bases.

An interesting change of the UV-absorbances with electrolyte concentration was observed for A 18 and T 18, as shown in Fig. 5. The molar extinction coefficient of A 18 decreased by about 7% at 0.09 mM, and that of T 18 about 10% at 0.16 mM. These concentrations may correspond to the critical micelle concentration, since the cmc observed from the surface tension measurements were about 0.1 mM for both A 18 and T 18.

Continuous variation mixing experiments in aqueous media have shown that A 12 and Poly U form a complex with a 1 : 1 stoichiometry in bases, and T 12 and

Structures

24, A12: $H_3C-N^+(CH_3)(C_{12}H_{25})-CH_2-CH_2-$[adenine]

25, T12: $H_3C-N^+(CH_3)(C_{12}H_{25})-CH_2-CH_2-$[thymine]

26, TH12: $H_3C-N^+(CH_3)(C_{12}H_{25})-CH_2-CH_2-$[theophylline]

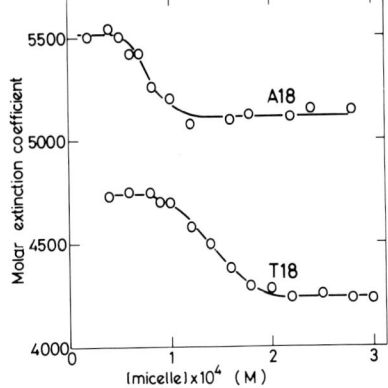

Fig. 5. Concentration dependence of the absorbance of A 18 and T 18 at 25 °C. A 18 at 260 nm, T 18 at 270 nm (Ref.[47])

Poly A form a complex with a 4:1 stoichiometry. The apparent hypochromicities of various mixtures are listed in Table 4. The mixtures of A 12 with Poly U and of T 12 with Poly A showed large hypochromicities compared with other mixtures, which suggests the importance of the hydrogen-bonding formation between complementary nucleic acid bases such as A-U and T-A.

A hypochromicity of about 10% was, however, observed between TH 12 and the polynucleotide. Furthermore, we observed a hypochromicity of 7% for the mix-

Table 4. Apparent hypochromicities of surfactant-type models of nucleic acid[47]

Mixtures	Solvent	λ_{max} (nm)	Hypochromicity (%)
A 12 + Poly U	H_2O	260	22
A 12 + Poly A	H_2O	260	12
A 12 + Poly A	H_2O	258	18
TH 12 + Poly U	H_2O	268	10
TH 12 + Poly A	H_2O	265	13

ture of dodecyldimethylbenzylammonium chloride and Poly U. Since theophylline and benzyl groups do not form a hydrogen bond with polynucleotide, these observations indicate that stacking-type hydrophobic forces are significant. The hypochromicity decreased upon addition of ethanol (Fig. 6), demonstrating the important contribution of hydrophobic interactions.

The important action of electrostatic forces between a cationic model and an anionic polynucleotide is clearly shown in Fig. 7. The hypochromicity sharply decreased with the ionic strength of the solution, which indicates that the base-base interactions between A 12 and Poly U supported by the electrostatic attractive forces are weakened by the shielding effects of added salts.

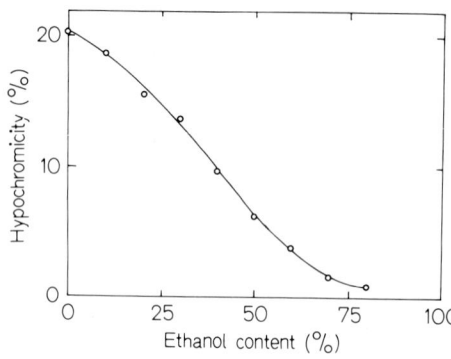

Fig. 6. Ethanol-content dependence of hypochromicity of Poly A + T 12 mixture at 25 °C. [Poly A + T 12] = 10^{-4} M (Ref.[47])

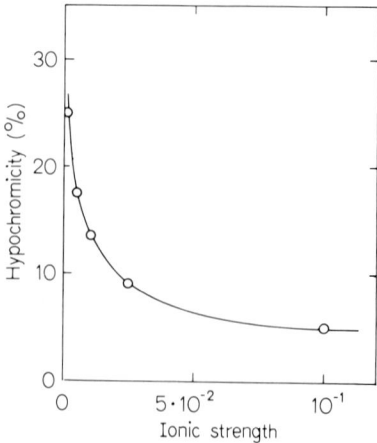

Fig. 7. Ionic-strength (KCl) dependence of hypochromicity of Poly U + A 12 mixture at 20 °C. [Poly U + A 12] = 10^{-4} M (Ref.[47])

2.5. Affinity Chromatography by Resin-Type Models

Affinity chromatographic procedures have been widely accepted for purification of biological macromolecules[48]. This technique is also one of the most convenient methods for analyses of interactions of solutes in eluates with support materials of the elution column. Tüchler[49] studied the base-base interactions using Amberlite ion-exchange resins containing nucleosides, and found that the affinities of *27* (A-Amberlite) and *28* (G-Amberlite) with bases were in the order $U > C$ and $C > U$, respectively, where U and C are uracil and cytosine, respectively. The authors proposed that there were specific interactions between complementary nucleosides.

Jones et al.[50, 51] synthesized a guanine derivative of cellulose, *29* (G-Cellulose), and found that the affinity of biosynthetic polynucleotides for the cellulose is in the order, poly $C >$ Poly A $>$ Poly U. They claimed that hydrogen-bonding between guanine residues and Poly C was probably taking place.

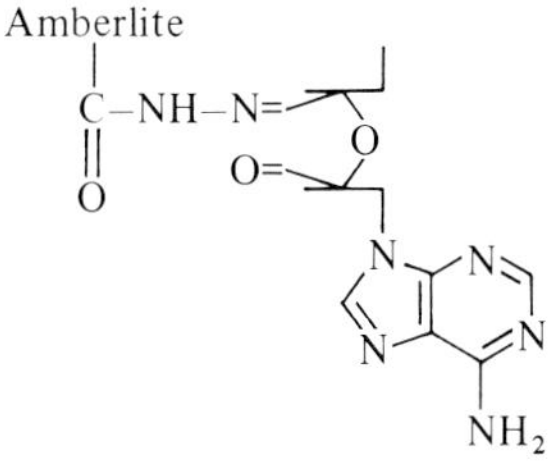

27, A–Amberlite

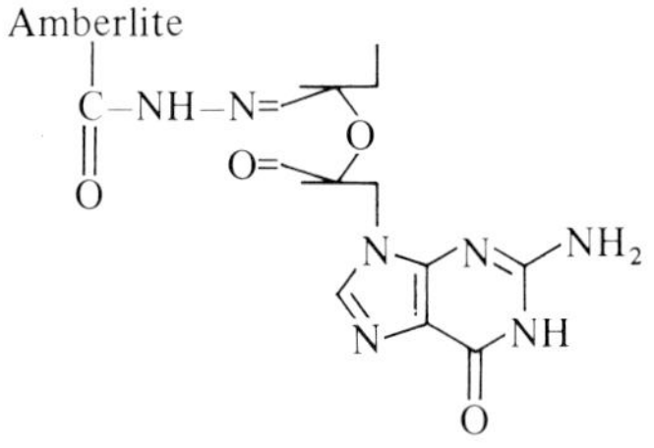

28, G–Amberlite

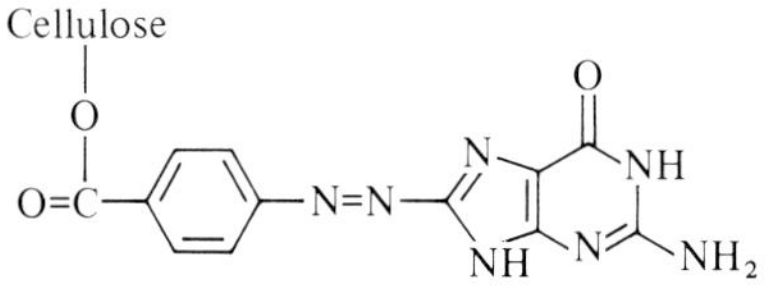

29, G–Cellulose

However, important contributions of hydrophobic stacking of bases were noted by Schott and Greber[52-55] using the data of the affinity chromatography of gels containing nucleosides, *30* (T-Gel), for example. Takemoto, Imoto, et al.[32] carried out the chromatographic separation of nucleic acid bases by using PU. Their results show that the separation is due exclusively to hydrophobic interactions.

We synthesized resin-type models with the Menschutkin reactions of poly-4-vinylpyridine-co-divinylbenzene or diethyl-aminoethylated cellulose with 9-(β-chloroethyl) adenine or 1-(β-chloroethyl) thymine[56]. A summary of the results is given in Fig. 8. The center of the bar indicates the fraction number corresponding to the elution maximum. Under our experimental conditions, a complete separation of the elution curves was not attained. This may be due to the fact that a comparatively small amount of resins was used and the experiments were conducted at a fairly high temperature. Looking at the results we may say that in water the nucleic acid bases are often paired by the complementary hydrogen-bonding, since the orders of $C < U$ and CMP $<$ UMP on *31* (ARPVP), and $U < A$ on TRPVP are interpreted by the order of the strength of hydrogen-bonding. AMP and UMP indicate adenosine-5′-monophosphate and uridine-5′-monophosphate.

Fig. 8. Summary of the affinity chromatography with RPVP, ARPVP, and TRPVP at 15 °C (Ref.[56])

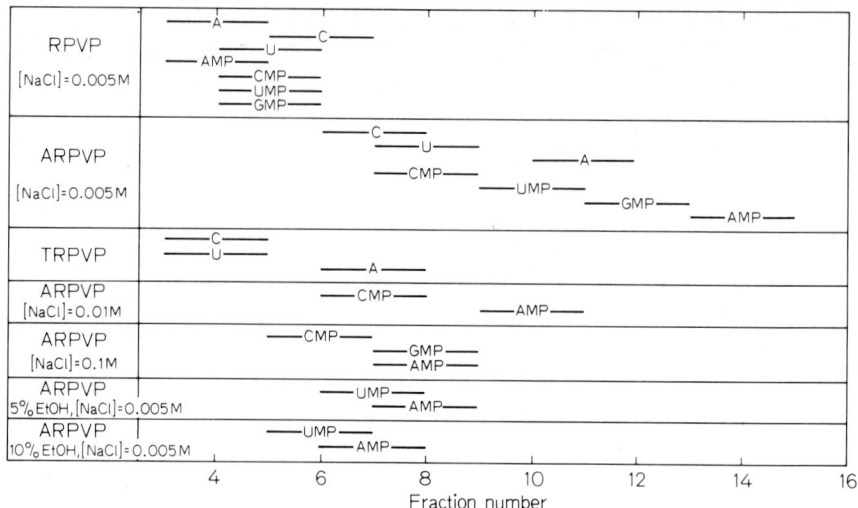

30, T–Gel

However, we found results suggesting the existence of the vertical base-stacking. The observed order of A > U on ARPVP (31) column could not be explained by the hydrogen-bonding but by the hydrophobic forces. It should be noted that even the order of A > U on the TRPVP column could be interpreted by the hydrophobic base-stacking strength, because the hydrophobicity of purine base was known to be stronger than that of pyrimidine.

Recently, Shimidzu et al.[57] studied the adsorption ability of cross-linked, ARPVP (31) resins toward mononucleotide in aqueous and pyridine media. They claimed that hydrogen-bonding was important between the resins and nucleotides under their conditions.

We also examined separation by 32 (A-DEAE cell) and T-DEAE cell. Uridine was more retarded by 32 (A-DEAE cell) than cytidine, and uridine more than adenosine. These results strongly suggested the existence of the hydrogen-bonding of A-U pairs. On the T-DEAE cell, adenosine was found to be more retarded than cytidine, and uridine more than cytidine, suggesting the complementary base-pairing. In conclusion, the hydrogen-bonding is more easily formed on *32* (A-DEAE cell) than on *31* (ARPVP). This may be ascribed partly to the hydrophobicity of DEAE cell resins being much weaker than that of RPVP.

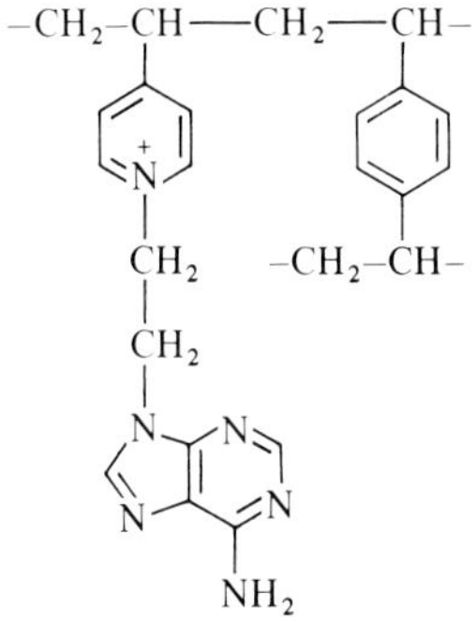

31, ARPVP 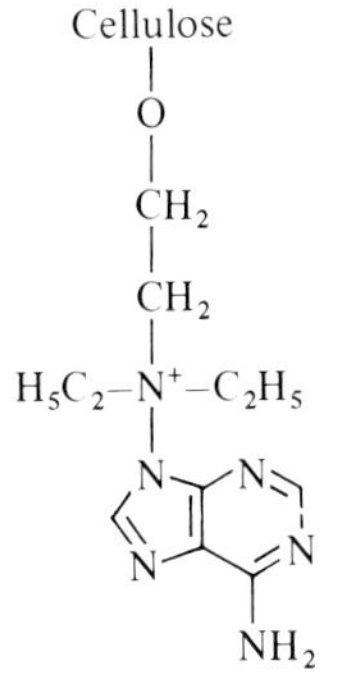   *32*, A–DEAE cell 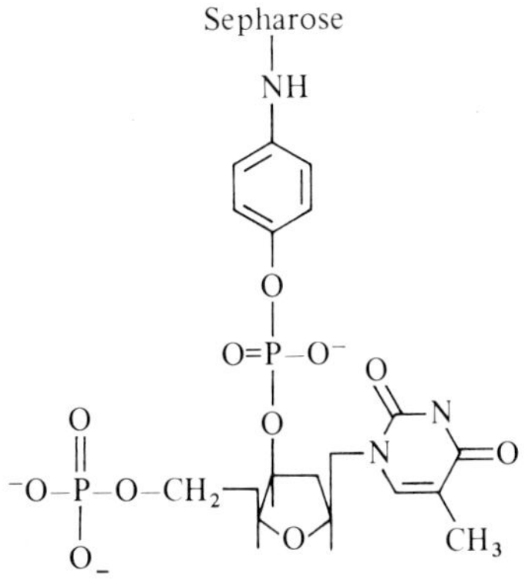   *33*, T–Sepharose

Anfinsen et al.[58] and Dunn and Chaiken[59, 60] utilized the affinity matrix, thymidine-3'-(p-Sepharoseamino-phenylphosphate)-5'-phosphate *33* (T-Sepharose) to examine nucleotide binding to staphylococcalnuclease. The authors found a close correspondence between the dissociation constants of matrix-bound and soluble thymidine-3'-(p-amino-phenylphosphate)-5'-phosphates. The results suggested that the interactions in the heterogeneous systems were quite similar to those occurring in solutions.

In relation to separation of nucleotides, Hoffman[61] found that adenine nucleotides interacted most strongly with cycloheptaamylose, presumably by inclusion of the base within the cavity of cyclodextrin. When epichlorohydrin-cross-linked cycloheptaamylose gel was used as a stationary phase for nucleic acid chromatography, adenine-containing compounds were retarded most strongly.

2.6. Template-Directed Polymerization of Nucleotides

With respect to template-directed syntheses, the syntheses of oligonucleotide using polynucleotide have already been attempted[62–69]. However, it has not necessarily been successful yet, since the binding constant of the nucleotide monomer with polynucleotide was not large, particularly in aqueous media.

By Naylor and Gilham[62], d (pT)$_6$ was condensed in the presence of Poly A with carbodiimide to give d-(pT)$_{12}$ with a yield of 5%. In the experiment of Shabarova and Prokofiev[67], d-(pA)$_2$ preactivated in the form of an amino acid amidate was condensed on Poly U to give products with a yield of 10%. Uesugi and Ts'o[68] studied the condensation of (2'MeIp)$_6$ or (2'MeIp)$_5$ in the presence of a Poly C template. The relative overall yield of the oligomer products was 43 to 71%.

The authors found that the yield of 30-mer (a product with 5–6 linkages) was not much smaller than that of 10-mer or 12-mer. These facts indicate that the stability of the complex between the oligonucleotides and the complementary template is the most important factor in determining the extent of the condensation. The strong influences of template polymer (Poly C) are demonstrated in Fig. 9, in which the elution profile is shown of the polymerization products of $(2'MeIp)_6$ in the presence of Poly C (B) and in their absence (A).

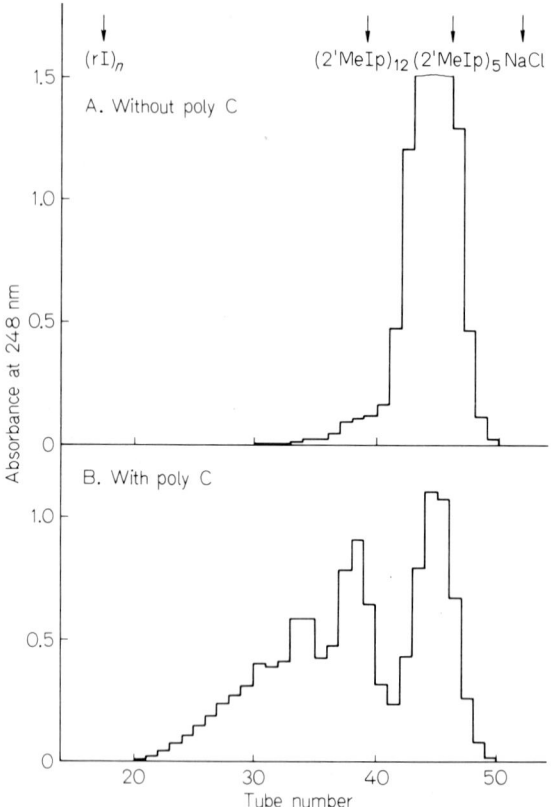

Fig. 9. The elution profile of the polymerization products of $(2'MeIp)_6$ in the presence of Poly C (B) and in the absence of poly C (A) (Ref.[68])

Simidzu et al.[70] carried out template-directed synthesis using various types of templates, e. g. neutral ones, *34* (PSt · TCT) and PSt · TCTC, and cationic ones, such as APVP, *35* (APVP · PSt), UPVP · PSt, ARPVP, APEI, and *36* (ARPEI). They obtained a large amount of purine-containing oligomers (dimer or trimer) as compared to pyrimidine ones by condensation of nucleotide mixture in the presence of the neutral templates containing pyrimidine bases, such as PSt · TCT and PSt · TCTC. The authors stated from the results that the existence of the sequence of the complementary base with the template was feasible. However, the above results can be explained in terms of the hydrophobic interactions, since the hydrophobicity of the purine monomer is higher than that of the pyrimidine one. Thus, the complementary base-pairing in the process of condensation can be more firmly established if purine templates are used.

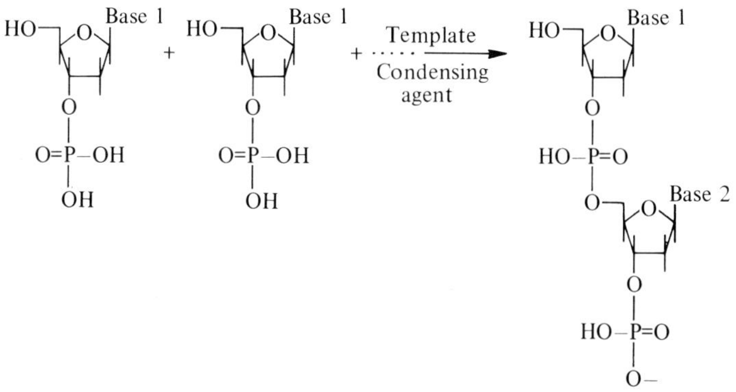

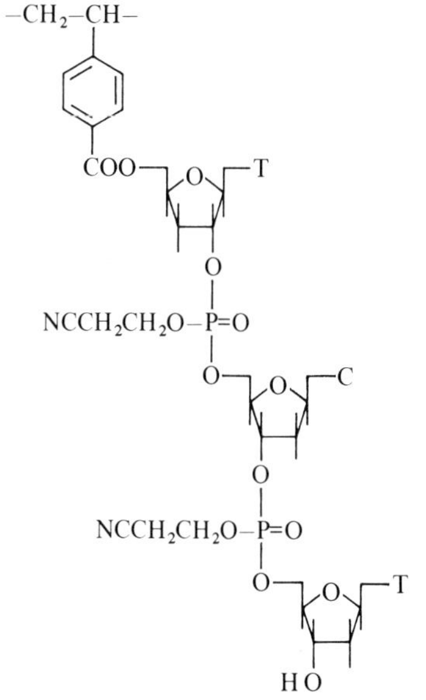

34, PSt · TCT

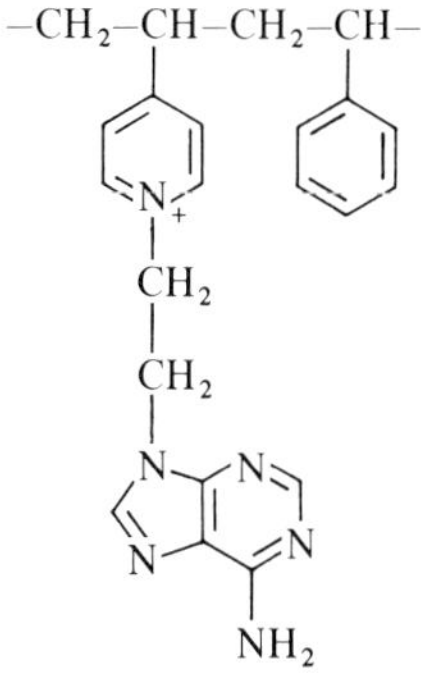

35, APVP · PSt

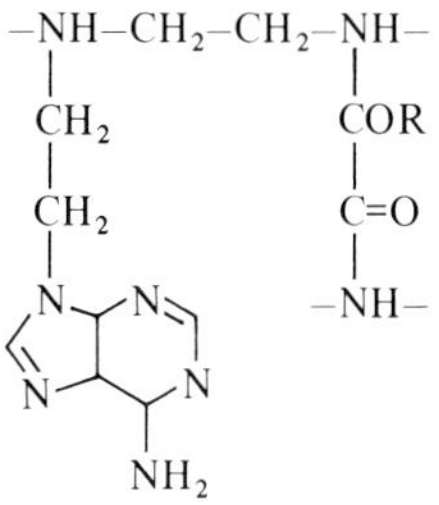

36, ARPEI

Shimidzu et al.[70] further studied polymerization by using cationic- and purine-containing templates, i. e. APVP, ARPVP, APEI, etc. The experimental results are compiled in Table 5. The complementary selectivity was defined by
$\{([pU] - [pA])/([pU] + [pA])\} \times 100\ (\%)$ for the equimolar binary monomer mixture

Table 5. Condensation of nucleotides in the presence of cationic templates[70]

Template	Monomers	Solvent	Complementary selectivity (%)
APVP	UMP + AMP	H_2O	0
APVP	UMP + AMP + CMP	H_2O	0
APEI	UMP + AMP	H_2O	0
APEI	TMP + AMP + CMP	H_2O	0
APVP · PSt	UMP + AMP	Pyridine	22
APVP · PSt	UMP + AMP + CMP	Pyridine	18
UPVP · PSt	UMP + AMP	Pyridine	14
UPVP · PSt	UMP + AMP + CMP	Pyridine	16
ARPVP	UMP + AMP	H_2O	40
ARPVP	UMP + AMP + CMP	H_2O	40
ARPVP	UMP + AMP	Pyridine	44
ARPVP	UMP + AMP + CMP	Pyridine	44
ARPEI	UMP + AMP	H_2O	0
ARPEI	UMP + AMP	Pyridine	24

of UMP and AMP, and by $\{([pU] - [pA] - [pC])/([pU] + [pA] + [pC])\} \times 100$ for the ternary mixture of UMP + AMP + CMP, where [pA], [pU], and [pC] denote the consumed amounts of the monomer by polymerization. As is clearly seen from the table, complementary oligomers were not obtained when water-soluble models (APVP, APEI) were used in aqueous media. On the other hand, a considerable amount of the complementary oligonucleotides was detected by using the water-insoluble templates, particularly in pyridine. These results seem to suggest that the heterogeneous systems are favorable to the formation of the complementary base-pairing. It should be remembered that a similar tendency was also noticed from the affinity chromatography (see Section 2.5.)

Now, it is interesting to note the ionic effects on renaturation kinetics of DNA. Wetmur et al.[71-75] investigated the kinetics of the renaturation of DNA in solution, which is an interionic process between anionic species. The renaturation rate was accelerated by addition of anionic polyelectrolytes, such as dextran sulfate, and this observation was accounted for by an excluded volume effect, which is certainly reasonable at high concentrations. Furthermore, the reaction rates were found to increase strongly with increasing ionic strength[71-79]. This effect was, quite recently, interpreted quantitatively by Manning[80] by using his theory on polyelectrolyte.

3. Esterase Models

To our knowledge, the first work on the use of synthetic polyelectrolytes as models of hydrolase was reported by Steinhardt and Fuggit[81]. They used polystyrenesulfonic acid, HPSt, for promotion of the acid hydrolyses of proteins instead of mineral acids. The polystyrenesulfonate anions were used because the macroions interact with both cationic substrates (proteins), and hydronium ions as catalysts of acid hydrolysis.

Since some polyelectrolytes could interact with nonionic substrates by other interactions, such as hydrophobic-, charge-transfer-, and hydrogen-bonding interactions, acid or alkaline hydrolyses of non-ionic compounds can be accelerated by the addition of polyelectrolyte. Furthermore, if we prepare macroions having hydrolytic active groups, the obtained polymer will be expected to promote the hydrolyses. Of course, the syntheses of the models of high efficiency, specificity, and regulation ability are not very easy. When we aim at obtaining good enzyme models, the polymer conformation and relative positions of various functional groups have to be controlled.

In the following sections, the present status of the investigation on esterase models will be described.

3.1. Polyelectrolyte-Catalyzing Acid Hydrolysis

As described above, Steinhardt and Fugitt[81] found that the catalytic activity of polystyrenesulfonic acid 37 (HPSt) was higher than that of mineral acid for the acid hydrolyses of proteins. Lawrence and Moore[82] investigated the hydrolysis rates of glycylpeptides more quantitatively by using cationic

$$H_3N^+-CHR-\overset{O}{\underset{\|}{C}}-NH-CHR'-\overset{O}{\underset{\|}{C}}-OH + H_3^+O \longrightarrow$$

$$H_3N^+-CHR-\overset{O}{\underset{\|}{C}}-OH + H_3N^+-CHR'-\overset{O}{\underset{\|}{C}}-OH \quad (3)$$

ion-exchange resins (Dowex-50). The rate with Dowex-50 was much higher than with HCl (see Table 6). They also determined the thermodynamic parameters, such as the enthalpy ($\Delta H^{\ddagger}$) and the entropy ($\Delta S^{\ddagger}$) of activation. Reaction (3) is assumed to be one of the interionic reactions between cationic reactants. Thus, the acceleration of the hydrolysis by macroanions is easily understood by the electrostatic interactions. However, quantitative discussion on the acceleration factor and on the thermodynamic

Table 6. Acid hydrolyses of glycylglycine with Dowex-50 and HCl at 54.5 °C[82]

Catalyst	$10^6 \times k_2$ (M^{-1}sec^{-1})	$\Delta H^{\ddagger}$ (kcal·mol^{-1})	$\Delta S^{\ddagger}$ (e.u.)
HCl	1.13	20.3	−24
Dowex-50	18.20	20.1	−19

parameters is rather difficult in relation to the present heterogeneous systems, because specific absorption of reactants and products on resins is significant.

Kern et al.[83, 84] carried out hydrolyses of peptides, such as glycylvarylalanine, glycyltryptophan and bovine serum albumin, in the presence of polyethylenesulfonic acid 38 (HPES) and 37 (HPSt) in the aqueous homogeneous systems. The observed acceleration factors ranged between 3 to 50.

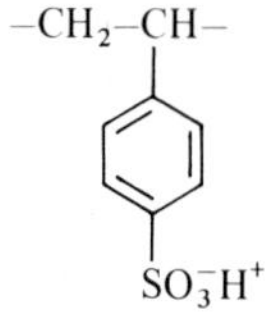

37, HPSt

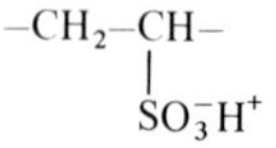

38, HPES

Yoshikawa and Kim[85, 86] hydrolyzed aminoacid esters (β-alaninemethylester, etc.) with 37 (HPSt) and 38 (HPES). They found an accelerating factor of ca. 10. The rate constants decreased with increasing degree of neutralization of the polymers.

Sakurada et al.[87] reported that aliphatic esters, such as methylacetate and n-butylacetate, were hydrolyzed more rapidly by more hydrophobic polymeric sulfonic acids. Furthermore, the rate of hydrolysis sharply decreased with the addition of acetone. The authors proposed that the rate enhancement was ascribed to the accumulation of the esters around the polymer by the hydrophobic interactions. Sakurada et al.[88] further examined the hydrolyses of polymeric esters by 37 (HPSt) or polystyrene partially sulfonated (HPSt*). The hydrolyses of acetylated polyvinylalcohol by 37 (HPSt) and HPSt* were more greatly accelerated with increasing degree of acetylation of the substrate, and with decreasing degree of sulfonation of the catalyst. These results also suggest an important contribution of hydrophobic interactions.

Some kinds of carbohydrates were hydrolyzed in the presence of polyelectrolytes. The hydrolysis of methyl-2-acetamide-2-dioxy-β-D-glucopyranoside with 37 (HPSt) was faster than HCl hydrolysis[83]. Furthermore, the catalytic activity of 37 (HPSt), for the hydrolyses of N-acetyllactoseamine or N, N'-diacetylthiobiose was 1.5 times higher than HCl[89]. On the other hand, the hydrolyses of the cationic ester, methyl-α-amino-2-dioxy-β-D-glucopyranoside hydrochloride by HPSt was 30 times more rapid than HCl[89], Diethylaminoethylated carbohydrates were hydrolyzed 20 times more efficiently by HPSt than HCl[90].

Arai and Ise[91] carried out the acid hydrolyses of dextrin in the presence of copolymers of vinylalcohol and vinylsulfuric acid, 39 (PVS · VA). Figure 10 shows

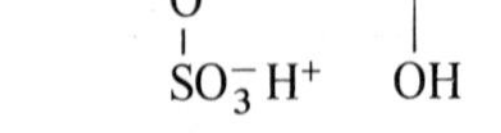

39, PVS-VA

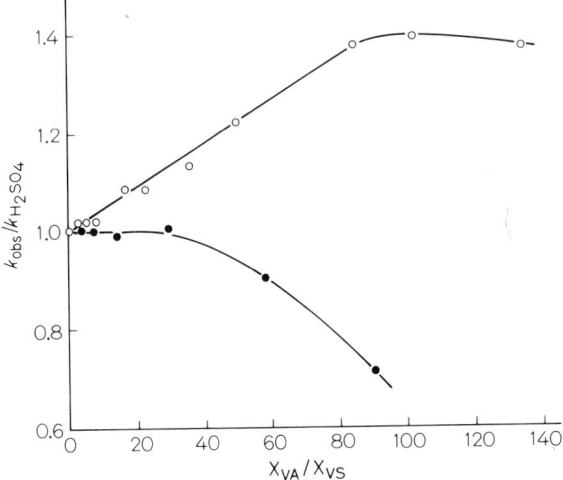

Fig. 10. Plots of the relative rate constants of the hydrolysis of dextrin vs. the mole ratio of vinylalcohol unit to vinylsulfonic acid unit. ○: PVS · VA, ●: Polyvinylalcohol + HPVS (Ref.[91])

the specific rate constant, k_{cat}, observed with the copolymers relative to that found with sulfuric acid. The relative rate constant increased with increasing vinylalcohol content in the copolymers at first and passed through a maximum. The authors explained the results as follows: the copolymers interact with the substrate by hydrogen-bonding and simultaneously with the protons, so that acceleration results. Arai et al.[92] also studied the hydrolyses of amylose and sucrose in the presence of copolymers of styrenesulfonic acid and acrylic acid 40 (PSt · AA).

Recently, Suzuki and Taniguchi[93] hydrolyzed n-butylacetate, ethylacetate, and methylacetate with HPSt* and 41 (PVA · B) (partially-o-benzalsulfonated polyvinylalcohol). The volume of activation, $\Delta V^{\ddagger}$, was obtained from the pressure dependence of reaction rates $[V^{\ddagger} = -kT(\partial \ln k/\partial P)]$. The $\Delta^{\ddagger}$ increased with increasing hydrophobicity of the substrate.

3.2. Polyelectrolyte-Catalyzing Alkaline Hydrolysis

Arcus and co-workers[94] studied alkaline hydrolyses of neutral and anionic aliphatic esters in the presence of poly (vinylbenzyltriethylammonium hydroxide), 42 (PVBzTEA), as a catalyst. Baumgartner and associate[95] also found enhanced alkaline hydrolyses of 43 (aspirin, anionic substrate), in the presence of 42 (PVBzTEA), in alkaline pH regions.

Recently, we conducted alkaline hydrolyses of p-nitro-phenylesters, i. e. 44 (PNPA), 45 (PNPPR), 46 (PNPV), 47 (PNPC), 48 (PNPL), and 49 (PNPP), using cationic polyelectrolytes 50 (C2PVP), 51 (C3PVP), 52 (C4PVP), 53 (BzPVP), and 54 (C16BzPVP) in a strong alkaline media using the stopped-flow technique[96].
In Fig. 11, the polyelectrolyte influences on the PNPP hydrolyses are shown. Though reaction (4) is not an interionic one, the present reaction was accelerated with the

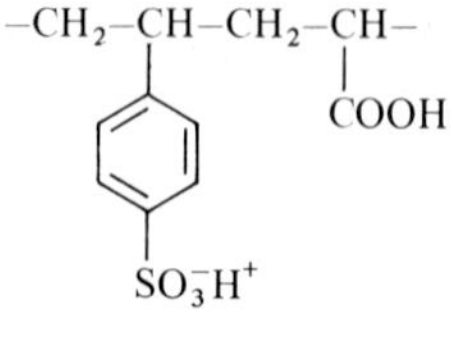

40, PSt·AA

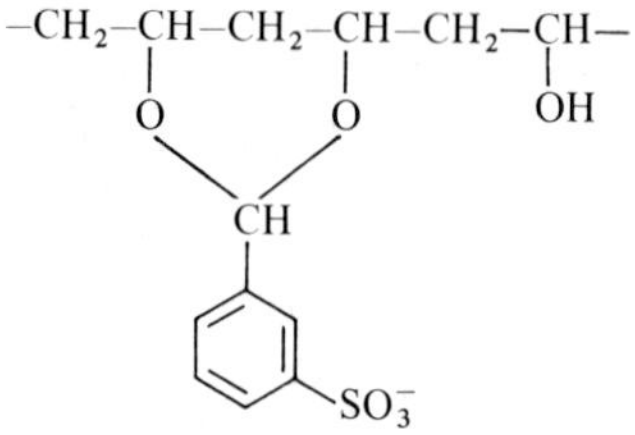

41, PVA·B

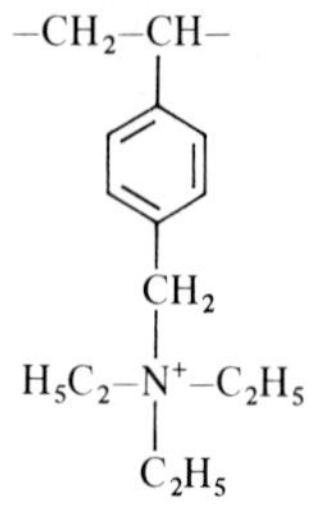

42, PVBzTEA

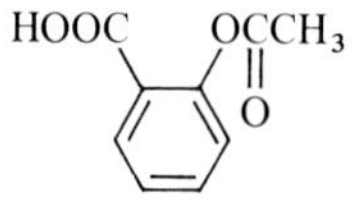

43, Aspirin

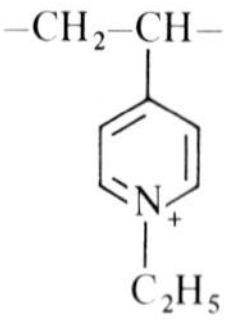

50, C2PVP 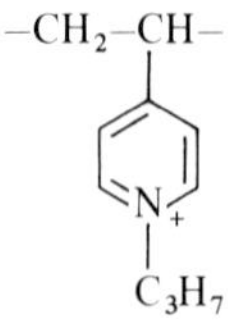 *51*, C3PVP 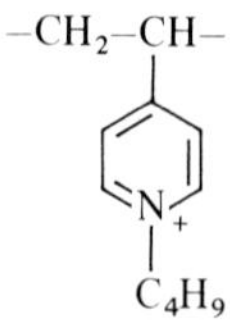 *52*, C4PVP 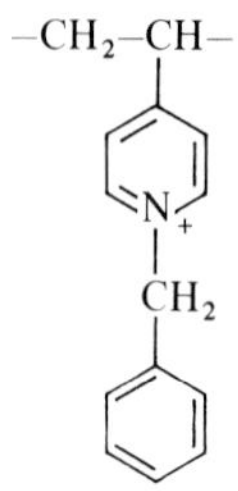 *53*, BzPVP 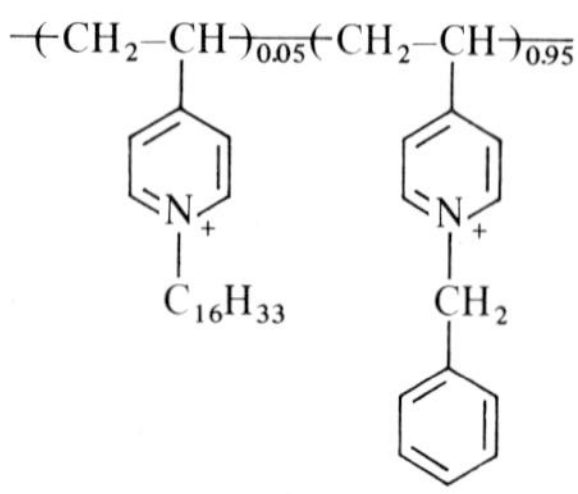 *54*, C16BzPVP

$$O_2N-C_6H_4-O-\overset{O}{\underset{\|}{C}}-OC_nH_{2n+1} + OH^- \longrightarrow O_2N-C_6H_4-OH + C_nH_{2n+1}COOH \quad (4)$$

n = 1	44, PNPA	n = 7	47, PNPC
2	45, PNPPR	11	48, PNPL
4	46, PNPV	15	49, PNPP

hydrophobic and cationic polyelectrolytes, and the strength of the acceleration increased with increasing hydrophobicity of esters and/or polyelectrolytes. This acceleration is clearly due to the hydrophobic interactions between polyelectrolytes and esters, and the electrostatic forces between polyelectrolytes and OH^-.

The three activation parameters, $\Delta G^\ddagger$, $\Delta H^\ddagger$, and $\Delta S^\ddagger$ decreased with polyelectrolyte addition. The decrease in $\Delta S^\ddagger$ suggests that the acceleration is due to the enthalpic loss. We recall that the acid hydrolyses of aliphatic esters with polymeric sulfonic acid was accompanied by decreases in $\Delta H^\ddagger$ and $\Delta S^\ddagger$ [97, 98].

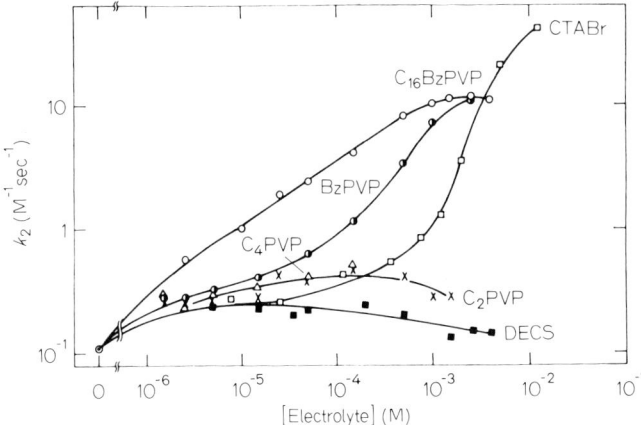

Fig. 11. Polyelectrolyte effect on the alkaline hydrolysis of PNPP at 30 °C in 30% ethanol-H_2O. [PNPP] = 4×10^{-5} M, [NaOH] = 1 mM (Ref.[96])

Cordes et al.[99] carried out alkaline hydrolyses of p-nitrophenylhexanoate 55 (PNPH) in the presence of poly-4-vinylpyridine partially quaternized with dodecylbromide and ethylbromide (QPVP). They also found that the polyelectrolytes are increasingly effective as catalysts with an increasing ratio of dodecyl to ethyl groups, and the hydrophobic interactions are important in determining the catalytic efficiency. They observed the inhibitory effects of several gegen-anions: fluoride ions are the weakest inhibitor, and nitrate is the strongest ($F^- < Cl^- < SO_4^{--} < NO_3^-$).

$O_2N-\underset{\text{55, PNPH}}{\underset{}{\bigcirc}}-O-\overset{O}{\underset{\|}{C}}-C_6H_{13}$

Furthermore, Cordes et al.[99] observed the saturation-type kinetics, strongly suggesting the formation of a complex between the polyelectrolyte and ester preceding bond cleavage reactions, as has been found for micellar catalysis[11, 101].

We also found the saturation kinetics for alkaline hydrolyses of 44 (PNPA), 3-nitro-4-acetoxybenzoic acid 56 (NABA), and 3-nitro-4-acetoxybenzenearsonic acid 57 (NABAA) in the presence of QPVP[102]. If ester-polymer complex formation occurs prior to the attack of OH^-, Eq. (5) holds, according to Bunton et al.[103], where K is the equilibrium association constant of polyelectrolyte (PE) and ester (S), and k_1 the first-order rate coefficients[103], PE, S, and P indicate the poly-

56, NABA   57, NABAA

electrolyte-ester complex, substrate and the product, respectively. We obtain Eq. (6) under the condition of [PE] >> [PE, S]. k_{obs}

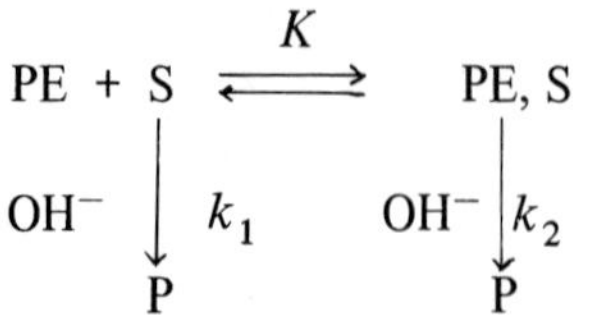

$$\begin{array}{c} PE + S \xrightleftharpoons{K} PE, S \\ OH^- \downarrow k_1 \quad\quad OH^- \downarrow k_2 \\ P \quad\quad\quad\quad P \end{array} \quad\quad (5)$$

is the observed first-order rate constant, n the number of

$$\frac{1}{k_{obs} - k_1} = \frac{1}{k_2 - k_1} + \frac{n}{k_2 - k_1} \frac{1}{K} \frac{1}{[PE]} \quad\quad (6)$$

binding sites of polyions forming complexes with ester molecules. The n-values for the univalent ester, NABA and the divalent, NABAA, are 1 and 2, respectively. For the neutral esters, we assumed $n = 1$. The values of K, k_1, and k_2 for various systems are compiled in Table 7.

Table 7. Parameters of Eq. (6) for the hydrolyses of phenylesters[102] (Ref.[102])

Ester	Catalyst	K (M^{-1})	k_1 (sec^{-1})	k_2 (sec^{-1})
NABA[a]	DECS	210	0.14	0.8
	C2PVP	260	0.14	1.2
	C4PVP	330	0.14	1.3
	C16BzPVP	330	0.14	3.2
NABA[b]	DECS	30	5.4×10^{-5}	7.2×10^{-4}
	C2PVP	90	5.4×10^{-5}	2.6×10^{-4}
	C16BzPVP	65	5.4×10^{-5}	3.3×10^{-4}
NABAA[c]	DECS	900	0.08	0.8
	C2PVP	1300	0.08	2.6
	C3PVP	200	0.8	6.3
PNPV[d]	C16BzPVP	520	0.20	1.0
PNPC[e]	BzPVP	260	0.11	0.22
PNPL[f]	BzPVP	640	0.016	0.21
	C16BzPVP	700	0.016	0.30
PNPP[g]	C4PVP	2400	1.1×10^{-4}	6.7×10^{-4}
	BzPVP	850	1.1×10^{-4}	0.022
	C16BzPVP	2600	1.1×10^{-4}	0.015

[a] at 25° in H$_2$O.
[b] at 30° in H$_2$O.
[c] at 35° in H$_2$O.
[d] at 30° in H$_2$O.
[e] at 25° in 15% EtOH-H$_2$O.
[f] at 25° in 22.5% EtOH-H$_2$O.
[g] at 30° in 30% EtOH-H$_2$O.

Recently, the quaternized poly-4-vinylpyridine, 50–54 (QPVP) was found to be an electron acceptor in the charge-transfer interactions[104]. Ishiwatari et al.[105] studied alkaline hydrolyses of p-nitrophenyl-3-indoleacetate 58 (p-NPIA) and N-(indole-3-acryloyl) imidazole 59 (IAI) (electron donor) in the presence of QPVP. The k_{obs} vs. polyelectrolyte concentration plots are shown in Fig. 12. As is seen in

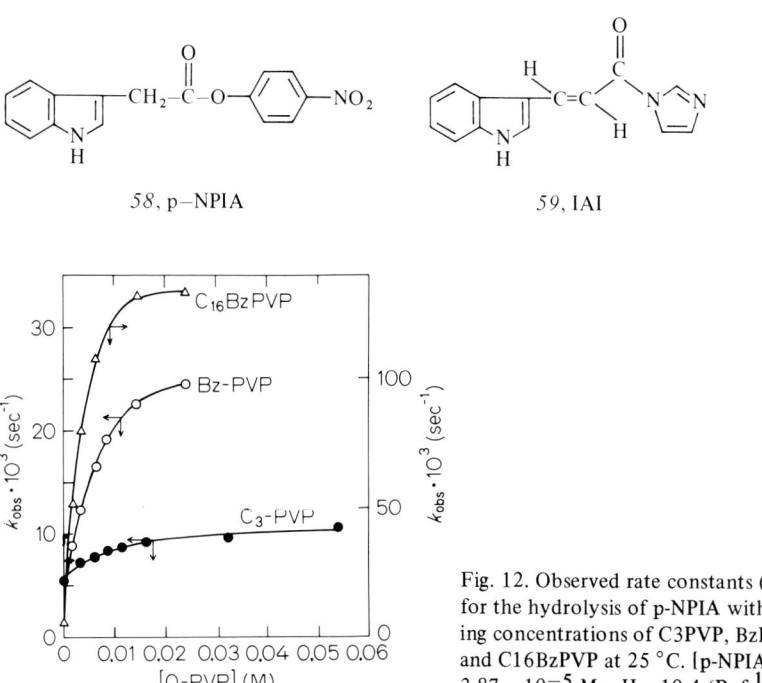

Fig. 12. Observed rate constants (K_{obs}) for the hydrolysis of p-NPIA with varying concentrations of C3PVP, BzPVP, and C16BzPVP at 25 °C. [p-NPIA] = 3.87 x 10^{-5} M, pH = 10.4 (Ref.[105])

this figure, saturation kinetics seem to be valid for the reaction. The parameters are compiled in Table 8 for the 59 (IAI) hydrolyses. Decreases of $\Delta H^{\ddagger}$ and $\Delta S^{\ddagger}$ by the addition of the polymer were found, and the contributions of the charge-transfer interactions were important, in addition to the electrostatic and hydrophobic interactions in the hydrolytic systems.

Table 8. Thermodynamic parameters for the hydrolyses of IAI at 25 °C[105]

Polymer	Concn. of Polymer (M)	$10^3 \times k_{obs}$ (sec^{-1})	$\Delta H^{\ddagger}$ (kcal · mol^{-1})	$\Delta S^{\ddagger}$ (e. u.)	$\Delta G^{\ddagger}$ (kcal · mol^{-1})
none	0	1.45	19.6	− 5.79	21.3
C3PVP	0.0345	1.81	16.0	−17.90	21.3
BzPVP	0.02	2.25	16.8	−14.30	21.2
C16BzPVP	0.005	4.17	13.3	−24.90	20.7

[IAI] = 3.96 x 10^{-5} M, borax buffer (pH = 10.4).

3.3. Polyelectrolytes Containing Nucleophile Groups

In this section, we restrict ourselves to the polyelectrolyte catalyst. A large number of investigations on imidazole-containing polymers has been reported. The readers are referred to other review articles[6, 106, 107].

So far, many kinds of nucleophiles active for hydrolysis such as imidazolyl-, amino-, pyridino-, carboxyl- and thiol-groups, have been used for preparation of hydrolase models. Overberger et al.[108, 109] prepared copolymers of vinylimidazole and acrylic acid 60 (PVIm · AA), by which the cationic substrate, 61 (ANTI), was hydrolyzed. This kind of copolymer is considered to be a model of acetylcholinesterase. With ANTI, the rate of the copolymer catalysis was higher than that of imidazole itself in the higher values of pH, as is seen in Table 9. In this work, important contributions of the electrostatic interactions are clear. The activity of the copolymer was not as high with the negatively charged and neutral substrates.

Overberger and associate[110] also prepared copolymers of 6(6)-vinylbenzimidazole with acrylic acid 62 (PBzIm · AA). In 40%-propanol water, the activities of the copolymers indicated a strong dependence on the carboxylate-benzimidazole-carboxylate triad. These effects were ascribed to strong electrostatic interactions between these sequences and substrates.

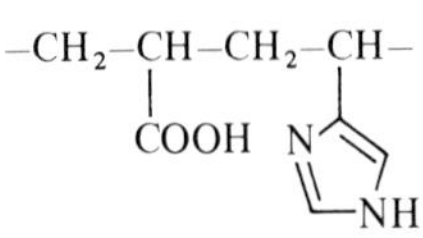

60, PVIm·AA

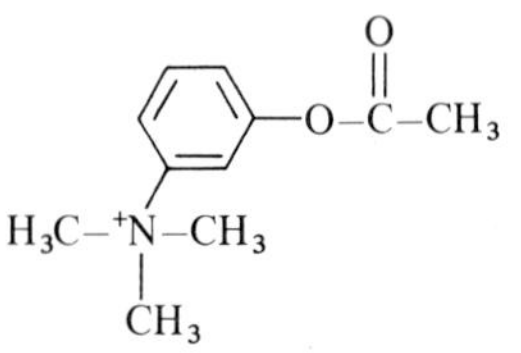

61, ANTI

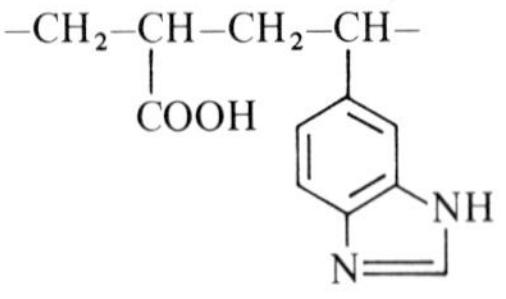

62, PBzIm·AA

Table 9. Catalytic rate constants of PVIm · AA and imidazole with ANTI[108]

pH	PVIm · AA k_{cat} ($M^{-1}min^{-1}$)	Imidazole
7.0	0.3	1.9
8.0	1.8	4.3
8.8	9.9	4.5
9.5	19.4	4.7

Shimidzu et al.[111] studied the catalytic activity of poly (4(5)-vinylimidazole-co-acrylic acid) 60 (PVIm · AA) in hydrolyses of 3-acetoxy-N-trimethylanilinium iodide 61 (ANTI) and p-nitrophenylacetate 44 (PNPA). The hydrolyses of ANTI followed the Michaelis-Menten-type kinetics, and that of PNPA followed the second-order kinetics. Substrate-binding with the copolymer was strongest at an imidazole content of 30 mol%. The authors concluded that the carboxylic acid moiety not only changed the fraction of the neutral imidazole moiety in the polymer, but also the nucleophilicity of the imidazole. For the function of carboxylic groups, Shimidzu et al.[112] further discussed the hydrolyses with a copolymer of vinylimidazole and γ-vinyl-γ-butyrolactone 63 (PVIm · VBI). The catalytic activity of the copolymer in PNPA and ANTI hydrolyses was found to be more than twice as high as that of random terpolymers containing imidazole, carboxylic acid, and hydroxyl moieties 64 (PVIm · A · AA). Similar studies using copolymers containing phenylimidazole unit and acrylate (or methacrylate) unit were carried out by Kunitake and Shinkai[113].

Cationic polyelectrolytes containing imidazole groups have been investigated by some researchers. Morawetz et al.[114] first found that a cationic polymer, poly (1-vinyl-3-ethylimidazolium iodide), 65 (PQMeIm), enhanced the hydrolyses of the negatively charged esters, i. e. NABA and 4-acetoxy-3-nitrobenzenesulfonate 66 (NABS). At intermediate pH, a large catalytic effect was observed and this was

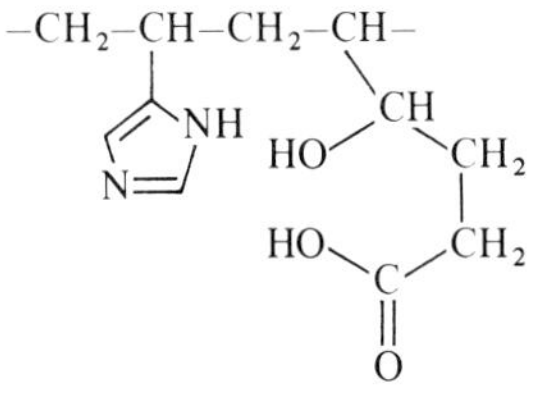

63, PVIm · VBL

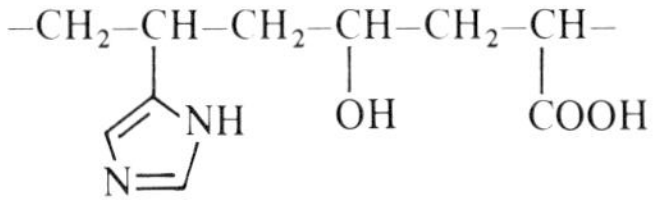

64, PVIm · A · AA

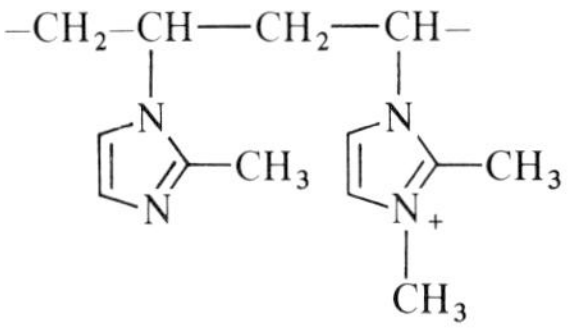

65, PQMeIm

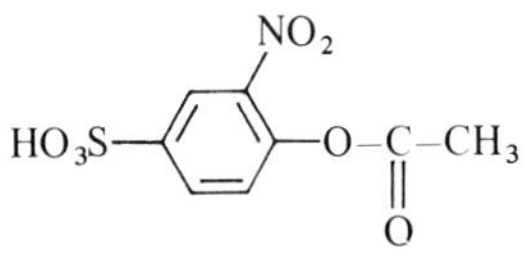

66, NABS

attributed to the enhanced susceptibility of the anionic ester to direct water attack in the vicinity of the macrocation. Kunitake and Shinkai studied the hydrolyses of 56 (NABA) with 65 (PQMeIM)[115]. They found the Michaelis-Menten kinetics for the hydrolyses. On the other hand, Overberger and Pacansky[116] did not find the saturation kinetics for the hydrolyses of NABA with the similar polymer, i. e. copolymer of vinylimidazole and vinyl-N-methylated-methylimidazole, 67 (PVIm · QMIm), in ethanol-aqueous media.

Kabanov et al. also reported that polyvinylpyridine quaternized with chloromethylimidazole showed solvolytic activity[117].

We recently synthesized cationic polymers containing imidazole [e. g. 68 (SZ811) and 69 (SZ11–3–3)] by reacting poly [N-(2,4-dinitrophenyl)-4-vinylpyridinium chloride] with histamine or histamine mixed with other amino derivatives [118]. The hydrolyses of neutral and anionic esters with the models followed saturation kinetics in alkaline media.

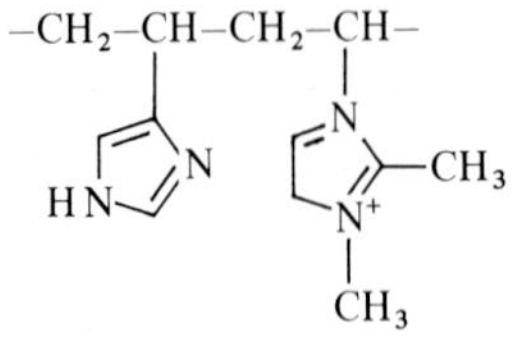

67, PVIm·QMIm

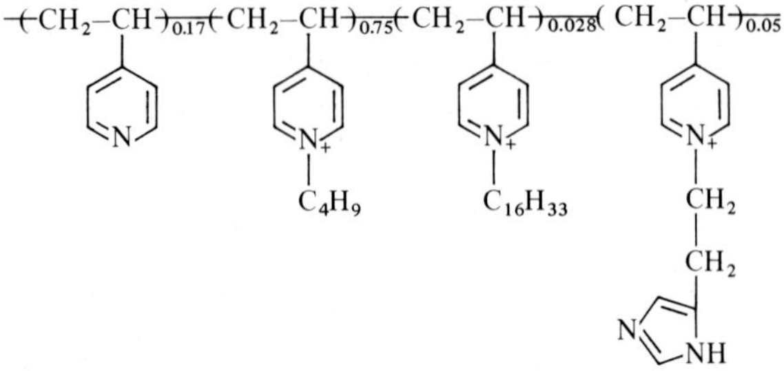

68, SZ·811

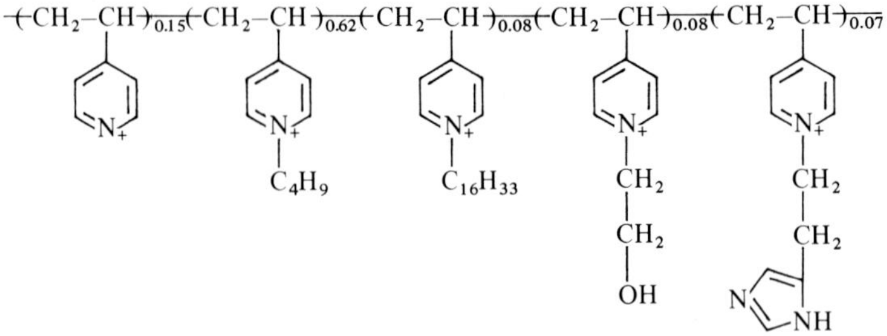

69, SZ–11–3–3

Klotz et al.[119, 120] asserted that polymers from polyethylenimine reacted with chloromethylimidazole and dodecyliodide were excellent models for hydrolase. 70 [PEI-D(10%)-Im(15%)] catalyzed the hydrolyses of PNPA 270 times faster than imidazole itself. The authors also found that lauroylated polyethylenimine containing hydroxamate and imidazole groups, 71 [PEI-HA(8%)-L(8%)-Im(6.6%)], had a high efficiency in hydrolyses, i. e. 310 times that of imidazole.

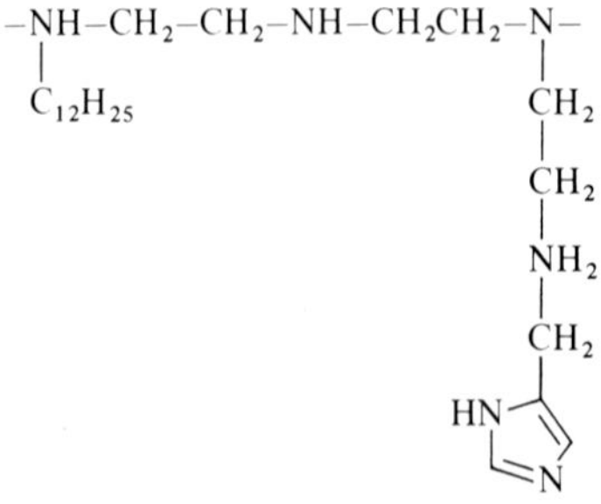

70, PEI–D(10%)–Im(15%)

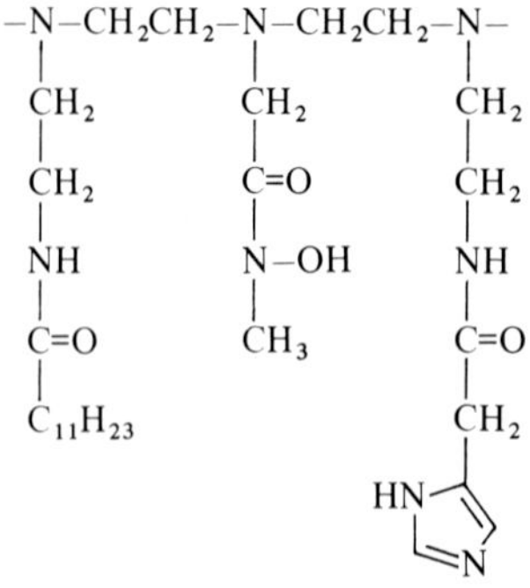

71, PEI–HA(8%)–L(8%)–Im(6.6%)

Klotz and associates[120] found that PEI-D-Im-type models catalyzed hydrolyses of 4-nitrocatechol sulfate by a two step mechanism resembling that of a natural enzyme. The polymer was reported to accelerate the rate by a factor of 10^{12} compared to imidazole itself, and of 10^2 compared to a type-11 A arylsulfatase enzyme.

$$O_2N-\underset{OH}{\underset{|}{\bigcirc}}-SO_3^- \longrightarrow O_2N-\underset{OH}{\underset{|}{\bigcirc}}-O^- + SO_3^{2-} \qquad (7)$$

Peptides containing histidyl groups were prepared and their catalytic activities discussed[121].

Pyridine groups show the nucleophilic catalytic activity. Letsinger and Saveride[122] hydrolyzed 2,4-dinitrophenyl-acetate, 72 (DNPA) and 3-nitro-4-acetoxybenzene-sulfonate, 66 (NABS) with partially quaternized poly-4-vinylpyridine (QPVP). They found that the rates for NABS increased as the degree of neutralization (α) was lowered and a maximum in the $k-\alpha$ plot appeared near $\alpha = 0.6$. For DNPA, such an enhancing effect was not detected.

$$O_2N-\underset{NO_2}{\underset{|}{\bigcirc}}-O-\overset{O}{\underset{\|}{C}}-CH_3$$

72, DNPA

Kabanov et al.[123] hydrolyzed 44 (PNPA) and 56 (NABA) in the presence of QPVP. They found that the hydrolysis rate decreased with increasing degree of quaternization in water. A bell-shaped dependence was observed for 56 (NABA) in alcohol-aqueous media (see Fig. 13). Recently, Cho and Morawetz discussed the hydrolyses of a polymer substrate by a pyridine-containing polymer[124].

Primary amino- and SH-groups are also nucleophiles and active for hydrolyses. Klotz et al.[125–127] investigated the activities of lauroyl-substituted polyethylenimine,

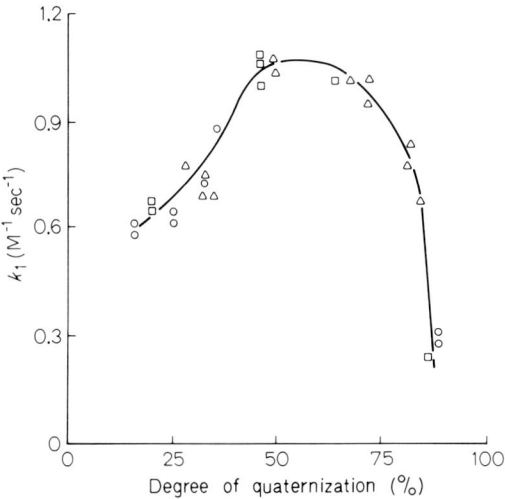

Fig. 13. Hydrolysis of NABA catalyzed by C2PVP(○), C4PVP(□), and C6PVP(△) in aqueous solution at 25 °C. [NABA] = 2.5×10^{-5} M (Ref.[123])

73 (L-PEI), mercaptoethylated polyethylenimine, 74 (SH-PEI) and both lauroyl-substituted and mercaptoethylated polyethylenimine, 75 (SH-L-PEI). These models showed an efficiency of 2 to 3 orders of magnitude larger than that of PEI itself

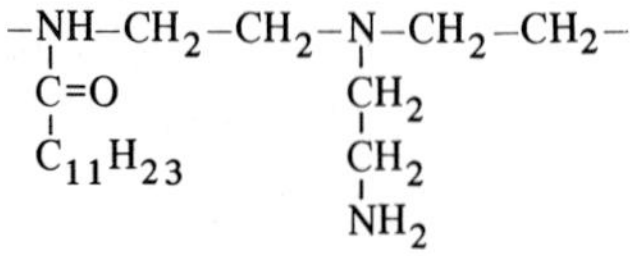

```
-NH-CH₂-CH₂-N-CH₂-CH₂-              -NH-CH₂-CH₂-N-CH₂-CH₂-
     |            |                       |            |
     C=O          CH₂                     CH₂          CH₂
     |            |                       |            |
     C₁₁H₂₃       CH₂                     CH₂          CH₂
                  |                                    |
                  NH₂                                  NH
                                                       |
                                                       CH₂
                                                       |
           73, L–PEI                                   CH₂
                                                       |
                                                       SH

                                              74, SH–PEI
```

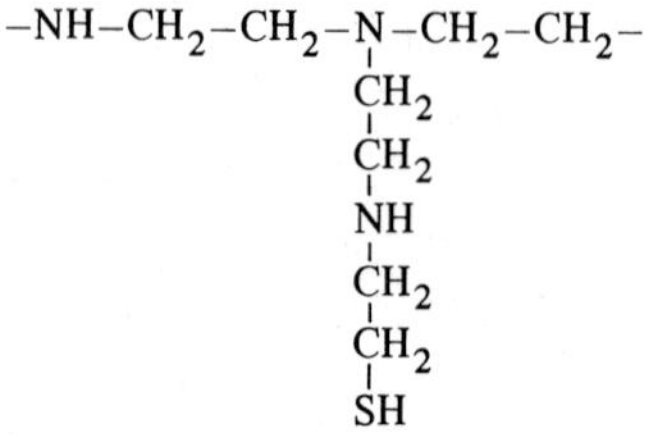

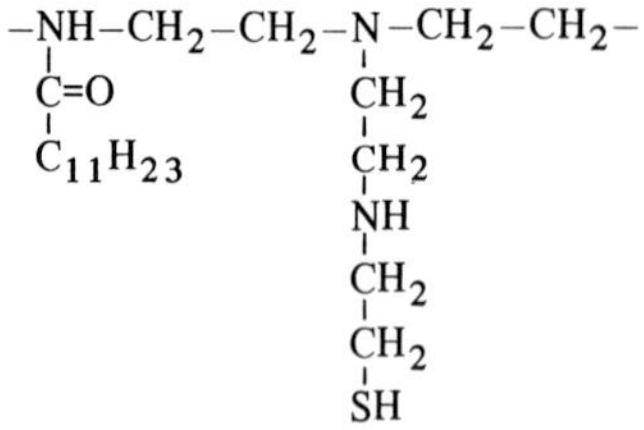

```
-NH-CH₂-CH₂-N-CH₂-CH₂-
     |            |
     C=O          CH₂
     |            |
     C₁₁H₂₃       CH₂
                  |
                  NH
                  |
                  CH₂
                  |
                  CH₂
                  |
                  SH

        75, SH–L–PEI
```

for the hydrolyses of p-nitrophenylester, 44 (PNPA). The order of the acceleration was PEI < L-PEI < SH-PEI < SH-L-PEI. The reactivity of amino groups of L-PEI was found to be enhanced by introducing hydrophobic lauroyl side groups; and the reactivity of an SH-group incorporated in a polymer was about 100 times greater than that of C_2H_5SH. It should be noted here that acylated amine of the models is stable against the hydrolytic cleavage. Therefore, the model compounds containing NH_2 or SH-groups act as acyltransfer agents but not as a catalyst.

Kabanov et al.[128] studied the hydrolyses of p-nitrophenylesters in the presence of linear and branched benzylated polyethylenimine, 76 (BzPEI). The catalytic site

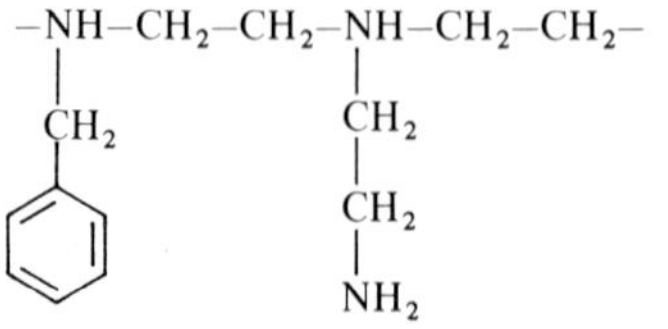

```
-NH-CH₂-CH₂-NH-CH₂-CH₂-
     |            |
     CH₂          CH₂
     |            |
    [Ph]          CH₂
                  |
                  NH₂

        76, BzPEI
```

was the benzylated tertiary amino group, which was ascertained by the isotope effect. The polymer, BzPEI, was 10^5 times more effective than dimethylbenzylamine for the hydrolyses of p-nitrophenyl-caproate, 47 (PNPC). There was no difference in the catalytic activity between linear and branched models.

The action of polyethylenimine on the hydrolyses of *43* (aspirin) was characteristic[95]; the reaction was substantially increased, passing through a maximum at pH = 7.8, where the rate constant was 1300 times greater than in the absence of the polyelectrolyte.

The hydrolyses of ADP and ATP were tried in the presence of low-molecular-weight analogs of polyethylenimine, *i. e.* ethylenediamine, *77* (ED), diethylenetriamine, *78* (DT), triethylenetetramine, *79* (TT), tetraethylenepentamine, *80* (TP) and pentaethylenehexamine[129]. These compounds accelerated the reaction several times. The degree of the acceleration increased with increasing degree of polymerization.

Recently, Kunitake and co-workers[130] reported a novel catalytic system consisting of C12-2PVP (poly-4-vinyl-pyridine quaternized with laurylbromide and ethylbromide) and N-methylmyristohydroxamic acid, *81* (MMHA). They observed

$$NH_2CH_2CH_2NH_2$$

77, ED

$$NH_2CH_2CH_2NHCH_2CH_2NH_2$$

78, DT

$$NH_2(CH_2CH_2NH)_3H$$

79, TT

$$NH_2(CH_2CH_2NH)_4H$$

80, TP

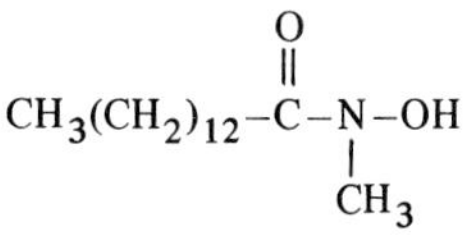

$$CH_3(CH_2)_{12}-\underset{\underset{CH_3}{|}}{\overset{\overset{O}{\|}}{C}}-N-OH$$

81, MMHA

a large enhancement of the nucleophilicity of the complexed hydroxamate anion toward several esters. For PNPA, the system enhanced the reaction 480 times more than with N-methyl-isobutyrohydroxamic acid only. This kind of complex would be one of the useful systems for achieving very high activity.

Kabanov *et al.*[131] found that a copolymer of 4-vinylpyridine and acroleinoxime, *82* (PP_{ox}), is a powerful catalyst for the hydrolyses of PNPA, NABA, and 3-nitro-4-trimethyl-acetoxybenzoic acid *83* (NTBA). The activity of the copolymer was 10^3 times higher than that of the low-molecular-weight oxime, iso-butyraldoxime. They proposed the cooperative activation of the oxime- and pyridine-groups in the vicinity of pyridinium cation groups of the copolymer.

Hayama *et al.*[132] discussed the catalytic effects of silver ion-polyacrylic acid systems toward the hydrolyses of 2,4-dinitrophenylvinylacetate *84* (DNPVA) by using the weak nucleophilicity of carboxylic groups and the change-transfer interactions between olefinic esters and silver ions[133]. Metal complexes of basic polyelectrolytes are also stimulating as esterase models. Hatano *et al.*[134, 135] reported that some copper(II)-poly-L-lysine complexes were active for the hydrolyses of amino acid esters, such as D- and L-phenylalanine methyl ester *85* (PAM). They

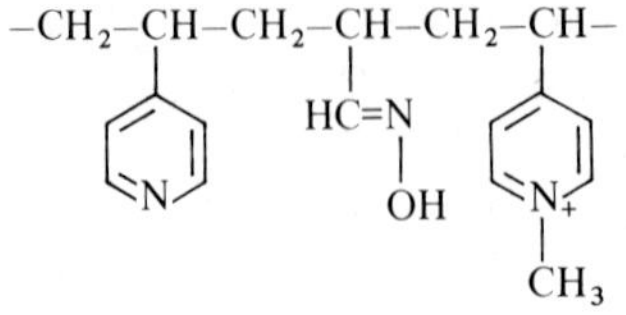

82, PPox

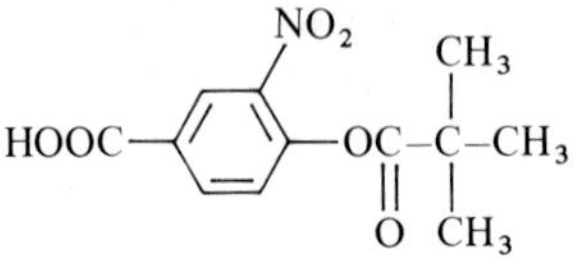

83, NTBA

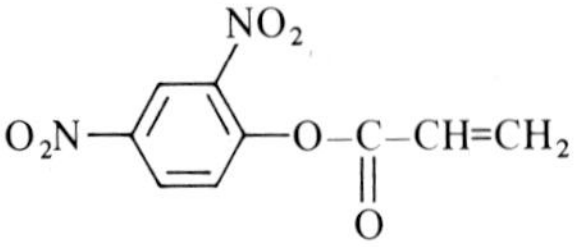

84, DNPVA

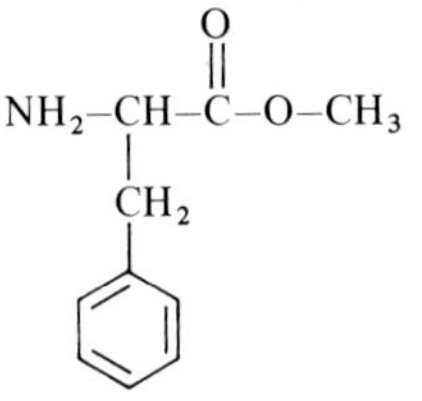

85, PAM

found a higher efficiency for the D-ester than for the L-isomer. This kind of stereospecificity may be intimately correlated to the conformation of the polymer.

3.4. Esterolytic Resin-Type Polyelectrolytes

Hydrolyses by heterogeneous systems are also important for their industrial utility and for obtaining selectivity. McGarvey and Kunin[136] published a review article of the relevant earlier work up to 1954. We quote here mainly the work published since then.

Studies by Deathrage et al.[137–139] revealed that most of dipeptides were hydrolyzed 100 times faster with cation exchange resins (Dowex-50) than with HCl. Deathrage et al.[139] also found that the entropy of activation was significantly less than in the case of hydrolysis of the same compounds by HCl, while the enthalpies of activation for the two cases were practically the same. While the entropy changes associated with catalysis by the cationic exchange resins remain obscure, presumably the mechanism of the catalysis follows that for homogeneous acids as described here later.

Hammet and collaborators[140, 141] studied in more detail the hydrolysis of aliphatic esters with a cation-exchange resins as catalyst. They found that replacement of 70% of the hydrogen ions in a crosslinked polystyrenesulfonic acid by cetyltrimethylammonium ions had a specifically favorable effect on the effectiveness of the remaining hydrogen ions for the hydrolysis of ethyl-n-hexanoate. From these findings, the important contributions of the hydrophobic forces, in addition to the electrostatic forces, is clearly demonstrated.

For the hydrolyses of neutral muco-polysaccharide, the activity of Amberlite IR-120 was 3 times higher than HCl 89,[142]. Painter et al. reported, interestingly enough, that both deglucoside bond and N-deacylation occurred in the presence of HCl, whereas N-deacylation did not occur in the presence of the resins[142]. As an example of the utility of the heterogeneous systems for obtaining selectivity of the hydrolysis reactions, the studies by Collins[143] are important. By the acid or alkaline hydrolyses of 1-benzoyl-1,2-dihydroquinaldamine (86), the cleavage of the 1-benzamide group and the hydrolyses of the 2-carbamoyl group occur simultaneously. However, the selective hydrolyses of compound (86) to 1-benzoyl-1,2-dihydroquinaldinic acid (87), was accomplished in the presence of Amberlite IR-120.

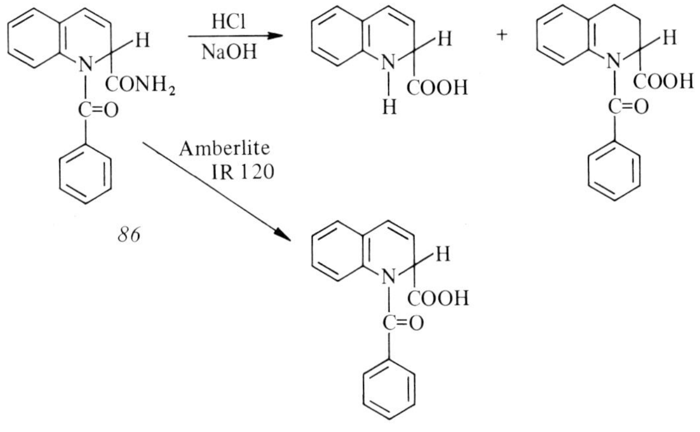

Cationic exchanger (crosslinked polystyrenesulfonic acid) was found to be very slightly more effective in the hydrolyses of cane suger than sulfuric acid[144].

Sakurada et al.[88, 145, 146] hydrolyzed aliphatic esters having various hydrophobicities with cation-exchange resins, Dowex 50 w, and found a close correlation between the accelerating factor and the degree of ester adsorption toward the resins.

We discussed more quantitatively the correlation of the adsorption with catalytic activity in the heterogeneous systems using poly(4-vinylpyridine-co-divinylbenzene) quaternized with alkylhalide [e. g. 88 (RCl 6PVP)] as a catalyst and p-nitrophenyl-valerate (PNPV) as a substrate[147].

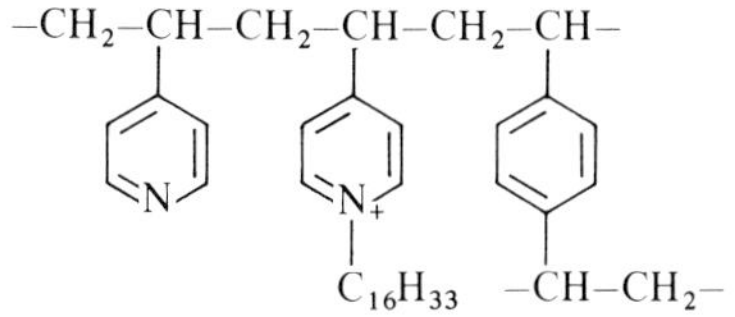

The concentration of the product absorbed in the resin phase and the real rate constant were determined by the measurements of the time dependencies of product formation in the bulk phase and of the quantities adsorbed to the resin of both the product and substrate by assuming the following reaction scheme:

$$R + S \underset{k_{-1}}{\overset{k_1}{\rightleftarrows}} RS \overset{k_2}{\to} RP \underset{k_{-3}}{\overset{k_3}{\rightleftarrows}} R + P \tag{9}$$

where R, S, and P denote the resin, substrate, and product, respectively. RS and RP are the complexes of the resin with the substrate and of the resin with the product. From Eq. (9) the rate of the product formation in the bulk phase is given by,

$$\frac{d[P]}{dt} = k_3[RP] - k_{-3}[R][P] \tag{10}$$

Thus, we obtain

$$[RP] = \frac{1}{k_3} \frac{d[P]}{dt} + \frac{k_{-3}}{k_3} [R][P] \tag{11}$$

$d[P]/dt$ is observable by spectrophotometric measurements. From the adsorption equilibrium of the product to the resin under the condition of $[R]_o \gg [S]_o$, we obtain

$$k_3 = \frac{[R]_o ([P]_o - [RP]_e)}{[RP]_e} \cdot k_{-3} \tag{12}$$

where $[X]_e$ denotes the concentration of X at adsorption equilibrium. Further, assuming the adsorption to be pseudo-first order, Eq. (13) holds

$$k_{ads} = k_3 + k_{-3} [R]_o \tag{13}$$

where k_{ads} is the pseudo-first order rate constant of adsorption. From Eqs. (12) and (13), we can determine the values of k_3 and $k_{-3} [R]_o$. Thus, by substituting these values into Eq. (9), [RP] is calculated, from which $[P]_{all} (= [P] + [RP])$ is calculable.

The real rate constant,

$$k_{cal} \left(= \frac{d[P]_{all}}{dt} \frac{1}{[S]_o} \right)$$

derived is shown in Fig. 14. Three curves show the data from the pH-stat method for comparison. It is surprising to note the excellent agreement of the spectroscopic data with the values from the pH-stat method for RC2PVP and *88* (RC16PVP).

As is clear from Fig. 14, the catalytic activity of quaternized resin is in the order, RC2PVP > RBzPVP > *88* (RC16PVP). This implies that the catalysis depends mainly on the hydroxide ion in the resin domain, because the degrees of quaternization change in the same order as that of the catalytic activity.

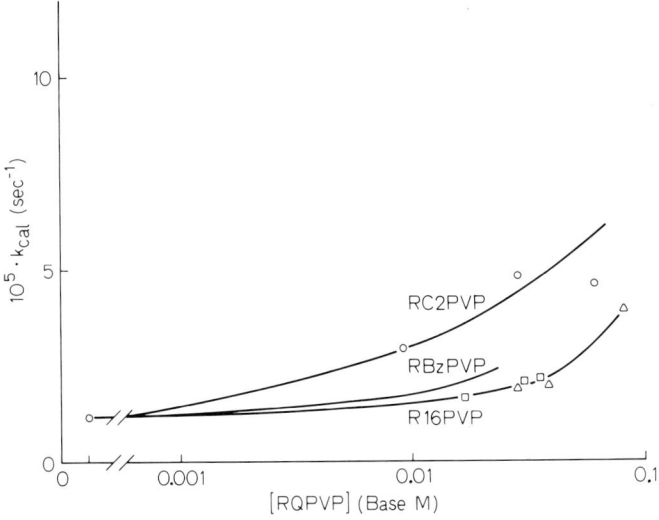

Fig. 14. Corrected rate constants (K_{cal}) of the hydrolysis of PNPV in the presence of RC2PVP(○), RBzPVP(△), and RC16PVP(□) at 25 °C (Ref.[147])

In our report[147], the thermodynamic parameters, $\Delta G^{\ddagger}$, $\Delta H^{\ddagger}$, and $\Delta S^{\ddagger}$ of the reaction in the presence of resins and in their absence were compared in detail. The results are shown in Table 10. As is clear in the table, both $\Delta H^{\ddagger}$ and $\Delta S^{\ddagger}$ decreased by the addition of *54* (C16BzPVP) in the homogeneous system and of *88* (R16PVP) in the heterogeneous system. This finding reveals that the kinetic mechanisms of the homogeneous and heterogeneous systems are similar, and the rate accelerations in both systems are due to the decrease of $\Delta H^{\ddagger}$, in accordance with the ones obtained earlier by Haskell and Hammet[148].

Affrossman[149] observed the accelerating effects of cation-exchange resins partially neutralized with AgOH toward the hydrolyses of allylacetate. π-Electron-binding interactions between Ag^+ and the double bond in allylacetate were proposed. π-Complex formations of olefine with Cu^+ and Ag^+ have already been reported[150, 151].

Table 10. Thermodynamic parameters of the PNPV hydrolyses in the homogeneous and heterogeneous systems at 25 °C[147]

Catalyst	Concn. of catalyst (mM)	$\Delta G^{\ddagger}$ (kcal · mol^{-1})	$\Delta H^{\ddagger}$ (kcal · mol^{-1})	$\Delta S^{\ddagger}$ (e. u.)
None (OH$^-$ only)	0	16.3	11.5	−16
C16BzPVP (homogeneous)	5.02	16.0	4.7	−38
RC16PVP (heterogeneous)	4.91	16.2	10.4	−20

[PNPV] = 0.464 mM, pH = 8.20, tris-HCl buffer, μ = 0.125, in 10 (v/v) % CH_3CN-H_2O mixture.

It is also stimulating to synthesize resins containing nucleophiles. We synthesized poly (4-vinylpyridine-co-divinyl-benzene quaternized with imidazole-4-ethylchloride *89* (RHisPVP)[147]. The catalytic activity of the resins was found to be significant for the hydrolyses of neutral *46* (PNPV). Murachi and Okumura[152] carried out the catalytic hydrolyses of *44* (PNPA) by using agarose-coupled glutathione. The experiments indicated a positive role of the bound glutathione molecules in accelerated hydrolyses of PNPA.

Arai and Ogiwara[153] prepared graft copolymer of vinylacetate on sulfonic-acid-type ion-exchange resins. Poly (vinylalcohol) grafted resins *90* (PVAG resins) obtained by saponification of the poly (vinylacetate)-grafted one had a catalytic activity toward the hydrolyses of amylose. This rate enhancement, though only slight, may be explained in terms of hydrogen-bonding between the substrate and the vinylalcohol unit.

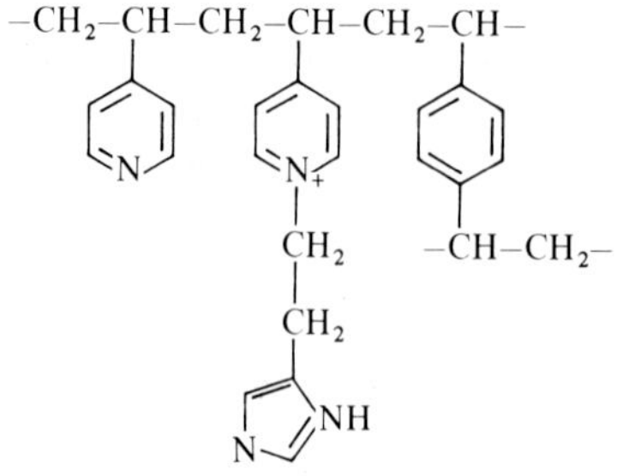

89, RHisPVP

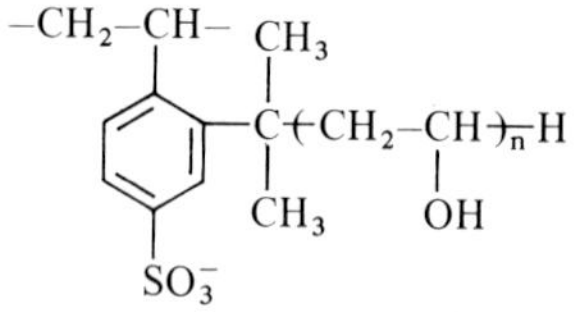

90, PVAG resins

3.5. Basic Aspects of Polyelectrolytes Retaining Esterase Activities

The knowledge of polyelectrolyte influences on the association constant or rate coefficient in ionic reactions is important when we attempt to design and synthesize polyelectrolyte models of esterase enzymes. It is now well known that interionic reactions between like-charged ionic reactants are largely enhanced by the addition of polyelectrolytes having charges opposite to the reactants, whereas reactions between oppositely charged molecules are retarded. These polyelectrolyte influences are surely attributable to electrostatic interactions between reactant ions and macroions. Qualitative interpretation of this effect has been given by some researchers. It was proposed that the rate-enhancing action of polyelectrolyte resulted from the "accumulation". The local concentration of ionic reactants around the polyelectrolyte catalyst is raised by the electrostatic forces, and hence the collision frequency of the reactants was increased[5, 154]. This interpretation is representative and seems to be qualitatively correct. However, it should be recalled that the acceleration often observed seems to be too great to be accounted for in terms of the "accumulation" effect only, because the reactants occupy definite exclusion volumes. The inadequacy of the accumulation factor was also noted by Morawetz and Gordimer[155] on the basis that the activation energy should not change by polyelectrolyte addition if this was the only factor, which is contrary to the observed fact. We thought it most pertinent to account for the catalytic influences of simple electrolytes and polyelectrolytes, and their other physico-chemical properties as well, in light of the common point of view or, in other words, by using the well-established physico-chemical principles.

It seemed to us that the concept of "primary salt effect" was worth consideration for the polyelectrolyte catalysis[156]. According to Brönsted[157] and Bjerrum[158], the rate constant of the reaction is accounted for in terms of the activated complex theory: $A + B \rightleftarrows X \rightarrow C + D$, X is the activated complex, C and D denote the product. The second-order rate constant, k_2, is given by

$$k_2 = k_{20} f_A f_B / f_X \tag{14}$$

where f is the single-ion activity coefficient, k_{20} is the limiting rate constant at zero ionic strength and given by kTK/h. K is the equlibrium constant between X and the reactants, h the Planck constant, k the Boltzmann constant and T the temperature.

If the activity coefficients are estimated from the Debye-Hückel theory in dilute regions of simple electrolyte systems, we have for aqueous solutions at 25 °C,

$$\log k_2 = \log(kT/h)K + 1.018\, Z_A Z_B I^{1/2} \tag{15}$$

where Z_A and Z_B are the valencies of the reactants, and I the ionic strength. The agreement between the theory and experiment has been excellent, and the validity of the theory has been firmly founded for simple electrolyte cases[158]. We extended the treatment for the polyelectrolyte-containing systems; the evaluation of f-values in the presence of polyelectrolytes were carried out by using Manning's theory on polyelectrolyte solutions[159–161].

The decelerating and accelerating actions of polyelectrolytes on the interionic reactions of $A^+ - B^-$ (or $A^- - B^+$) and $A^{m+} - B^{m+}$ (or $A^{m-} - B^{m-}$) types were found to be formulated by the Eqs. (16) and (17), respectively.

$$\log(k_2/k_2^*) = \log\left(\frac{1 + \xi^{-1} X}{X + 1}\right) - \frac{X}{X + 2\xi} \tag{16}$$

where $X = n_2/n_s$, n_e, and n_s are the concentrations of macroanions and the simple ions (reactant ions). k_2 and k_2^* denote the reaction rates in the presence of polyelectrolytes and in their absence.

$$k_2/k_{2R} = \left[X - \frac{1}{m}\left(1 - \frac{1}{m\xi}\right)X^2\right] \Big/ \left[X_R - \frac{1}{m}\left(1 - \frac{1}{m\xi}\right)X_R^2\right] \tag{17}$$

where k_{2R} is the rate constant at a reference concentration of polyelectrolyte (usually at the lowest polymer concentration employed). $X = n_e / (n_A + n_B)$. Excellent agreement between the theories and experiments was obtained. The comparisons for the ammonium-cyanate-urea conversion[162] and Hg^{2+}-induced aquation of $Co(NH_3)_5Br^{2+}$ in the presence of polyelectrolytes[163] are shown in Figs. 15 and 16, respectively. Reasonable agreement between the theories and experiments was also obtained for other types of interionic reactions, i. e. $A^{m+} - B^{n+}$-type reactions ($m > n$, $m > 2$) in the presence of macrocations or macroanions and $A^{z+} - B^-$ type reactions with macrocations.

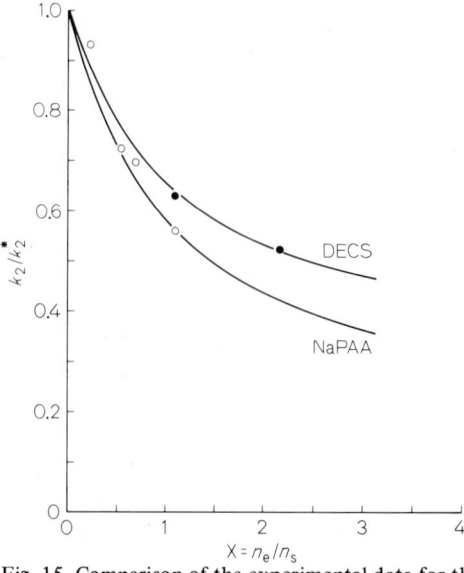

Fig. 15. Comparison of the experimental data for the polyelectrolyte effect on k_2 of NH_4^+-OCN^- reaction at 50 °C with the theoretical curves obtained from eq. 16. ○: NaPAA added, $[NH_4OCN]$ = 0.0205 M, ○: DECS added, $[NH_4OCN]$ = 0.1025 M (Ref.[159])

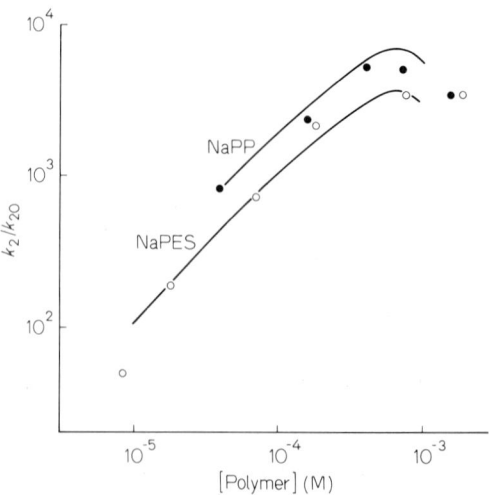

Fig. 16. The catalytic influence of NaPES and NaPP on the Hg^{2+}– induced aquation. $[Co(NH_3)_5Br^{2+}]$ = 6 × 10^{-5} M, $[Hg^{2+}]$ = 5 × 10^{-4} M. ○: obs. for NaPES, ●: obs. for NaPP, ——: calc. from Eq. (17), the reference concentration was taken to be 1.8 × 10^{-5} M (Ref.[160])

Thus, it may be concluded that the Brönsted-Bjerrum-Manning treatment, or the extended primary salt effect, gives a fairly satisfactory account of the general features of the electrostatic acceleration or deceleration influences of polyelectrolytes on various types of interionic reactions. It should be noted, of course, that the contribution of other interactions, such as hydrophobic[96, 99, 102, 164–167], hydrogen-bonding, and charge-transfer forces, which were also very significant, was neglected in Manning's theory, though the Brönsted-Bjerrum approach itself seems applicable to systems with these interactions, if the activity coefficient terms are evaluated experimentally.

According to Eq. (14), a great acceleration or deceleration can be observed for interionic reactions in the presence of ionic species, because f_A, f_B, and f_X differ largely from the ideal values in such systems as a consequence of most influential electrostatic interactions. Equation (14) also applies to molecule-molecule, or ion-molecule reactions, but the f-values should be generally close to the ideal value, giving rise to rather small catalytic influence (or $k_2/k_{20} \approx 1$). This situation leads us to the conclusion that reactants and "catalyst" should carry electric charges of the opposite sign if pronounced catalysis is to be expected. Furthermore, we believe that the true rate constant of complicated reactions (e. g. enzyme reactions), which involve both electrostatic and non-electrostatic interactions, can be evaluated only when the contribution of the extended primary salt effect is substracted (or, to be exact, corrected) from the observed rate data.

One of the most typical examples demonstrating the importance of the hydrophobic contribution may be alkaline-fading reactions of triphenylmethane dyes, which take place between dye cations and hydroxyl ions to form carbinols[167]. Since the dyes contain hydrophobic moieties, the fading reactions, which should be decelerated by polyelectrolytes according to the electrostatic model, can even be accelerated by hydrophobic polyelectrolytes. As shown in Fig. 17, the addition of

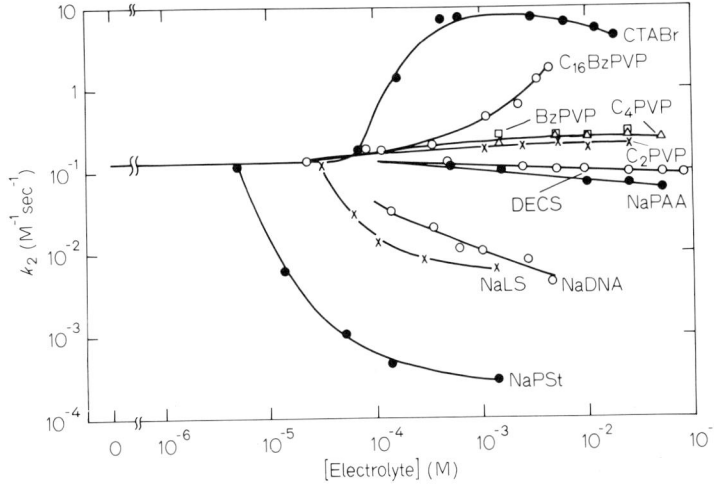

Fig. 17. Catalytic action of polyelectrolytes on the fading reaction of ethylviolet (EV) at 30 °C. [EV] = 1.05 x 10^{-5} M, [OH$^-$] = 1.05 x 10^{-2} M (Ref.[167])

hydrophobic cationic polymers (QPVP) and cationic surfactant (CTABr) enhanced the reaction, whereas anionic ones retarded it. It is reasonable to ascribe the observed acceleration to hydrophobic interactions between the substrate dye cations and the polyelectrolytes and to electrostatic attractive forces between OH^- and the polycations. The most important aspect in the observed acceleration is that the hydrophobic interactions overwhelmed the electrostatic repulsive forces between the dye cations and the polycations in the present reaction system. The strong retardation by 37 (NaPSt) is due to the cooperative contribution of the hydrophobic and electrostatic interactions between the polyanions and dye cations. Other polyanions, such as sodium polyacrylate, 91 (NaPAA), which lack the hydrophobic groups, decelerated the reactions much more moderately than NaPSt, because only the electrostatic interactions are operating, In Fig. 17, NaDNA and 92 (DECS) denote the sodium salt of salmon sperm DNA and a copolymer of diethyldiallyl ammonium chloride and sulfur dioxide, respectively. NaLS is sodium laurylsulfate. It should be remembered that the theoretical interpretation of the hydrophobic effects is difficult and not yet been achieved.

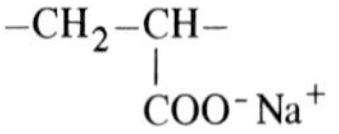

91, NaPAA

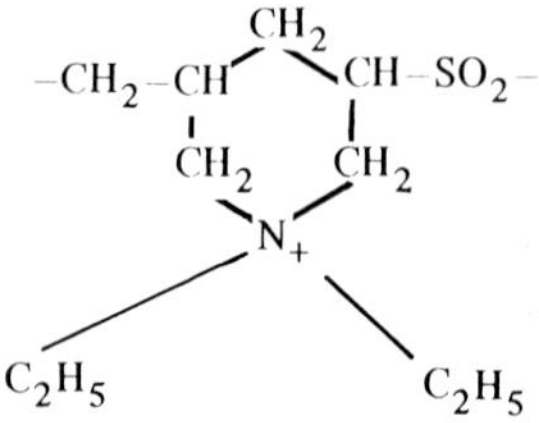

92, DECS

When esterase models are designed, several important and fundamental problems have to be solved. Systematic studies on other interactions, such as hydrogen-bonding and charge-transfer type forces have not been fully performed. Furthermore, various cooperative actions between different kinds of interactions, e. g. the correlation between the attraction of substrate and repulsion of a product by a polyelectrolyte catalyst, has not yet been carried.

4. Conclusion

In the foregoing chapters, we reviewed the present status of recent investigations on the synthetic polyelectrolyte as models of nucleic acids and esterases. Concerning nucleic acid models, the main efforts have hitherto been focused on the synthesis of vinyl-type polyelectrolytes containing nucleic acid bases, pentoses, and/or dissociable ionic groups. These polymers are of value for the elucidation of the independent roles of the groups in the function of nucleic acids. However, they have no defined molecular weights, stereoregular conformations, or well-defined chemical sequences. The syntheses of the model compounds satisfied with these conditions are certainly very important. Furthermore, the systematic investigations on the mechanism of the association with other biological substances and with themselves are still quite few. At present, we may safely say that the result obtained from the studies carried out so far show the important contributions of electrostatic and hydrophobic interactions besides hydrogen-bonding forces between nucleic acid bases. Base-base pairs due to hydrogen-bonding were found not to be formed without the assistance of the other interactions for almost all models obtained. At any rate, investigations in this field have just begun.

As enzyme models, polyelectrolytes should exhibit high efficiency, high selectivity, and regulatory function, if they are to be regarded as superior models[168]. Pertaining to the high efficiency, our purpose was at least partly attained in some cases. Large accelerating factors, $10^4 - 10^7$, have been found for interionic reactions in the presence of polyelectrolytes[1-3]. However, synthetic polyelectrolytes have been almost entirely incapable with regard to the specificity and the regulatory function. Elaborate syntheses of polyelectrolytes, which have defined stereoregularities, sequences, and much more effective functional groups reacting with substrates, are certainly desirable, though the task is obviously formidable.

Acknowledgment. We gratefully acknowledge the invitation and encouragement received from Professor Seizo Okamura to write this article. His critical reading of the manuscript is also deeply appreciated. Our thanks are due to all collaborators.

5. References

1) Ise, N.: Adv. Polymer Sci. 7, 536 (1971)
2) Morawetz, H.: Macromolecules in solution. 2nd edit. New York: Interscience 1975
3) Ise, N.: Polyelectrolyte and their applications. Selegny, E. and Rembaum, A. (eds.) Dordrecht-Holland: D. Reidel 1975, p. 71
4) Takemoto, K.: J. Macromol. Sci. Part. C 5, 29 (1970)
5) Sakurada, I.: Pure Appl. Chem. 15, 453 (1967)
6) Overberger, C. G., Salamone, J. C.: Accounts Chem. Res. 2, 217 (1969)
7) Morawetz, H.: Accounts Chem. Res. 2, 354 (1970)
8) Fendler, J., Fendler, E.: Catalysis in micellar and macromolecular systems. New York: Academic Press 1975
9) Seita, T., Yamaguchi, K., Kinoshita, M., Imoto, M.: Bull. Chem. Soc. Japan 45, 926 (1972)
10) Seita, T., Yamaguchi, K., Kinoshita, M., Imoto, M.: Makromol. Chem. 154, 255 (1972)
11) Seita, T., Yamaguchi, K., Kinoshita, M., Imoto, M.: Makromol. Chem. 154, 263 (1972)
12) Seita, T., Kinoshita, M., Imoto, M.: Makromol. Chem. 164, 7, 345 (1973)
13) Lipsett, M. N., Heppel, L. A., Bradley, D. F.: J. Biol. Chem. 236, 857 (1961)
14) Seita, T., Yamauchi, K., Kinoshita, M., Imoto, M.: Makromol. Chem. 164, 15 (1973)
15) Okubo, T., Ise, N.: Polymer J. 4, 675 (1973)
16) Okubo, T., Ban, K., Ise, N.: Makromol. Chem. 175, 49 (1974)
17) Strauss, U. P., Gershfeld, N. L.: J. Phys. Chem. 58, 747 (1954)
18) Fuoss, R. M., Strauss, U. P.: J. Polymer Sci. 3, 602 (1948)
19) Ts'o, P. O. P., Melvin, I. S., Olson, A. C.: J. Amer. Chem. Soc. 85, 1289 (1963)
20) Shimidzu, T., Murakami, A.: 23rd Polymer Symposium Preprint p. 9 Tokyo (1974)
21) Kondo, K., Iwasaki, H., Nakatani, K., Ueda, N., Takemoto, K., Imoto, K.: Makromol. Chem. 125, 42 (1969)
22) Kaye, H.: J. Amer. Chem. Soc. 92, 5777 (1970)
23) Pitha, P. M., Pitha, J.: Biopolymers 9, 965 (1970)
24) Michelson, A. M.: The chemistry of nucleosides and nucleotides. London: Academic Press 1963
25) Scheraga, H. A., Nemethy, G., Steinberg, I. Z.: J. Biol. Chem. 237, 2506 (1962)
26) Pitha, J., Ts'o, P. O. P.: J. Org. Chem. 33, 1314 (1968)
27) Pitha, J., Pitha, P. M., Ts'o, P. O. P.: Biochim. Biophys. Acta 204, 39 (1970)
28) Pitha, J., Michelson, A.: Biochim. Biophys. Acta 204, 381 (1970)
29) Pitha, J., Pitha, P. M., Stuart, E.: Biochemistry 10, 4595 (1971)
30) Pitha, J., Pitha, P. M.: Science 172, 1146 (1971)
31) Kondo, K., Iwasaki, H., Ueda, N., Takemoto, K., Imoto, M.: Makromol. Chem. 125, 298 (1969)
32) Ueda, N., Nakatani, K., Kondo, K., Takemoto, K., Imoto, M.: Makromol. Chem. 134, 305 (1970)
33) Takemoto, K., Kawakubo, F., Kondo, K.: Bull. Chem. Soc. Japan 44, 1718 (1971)
34) Kaye, H.: J. Polymer Sci B 7, 1 (1969)
35) Kaye, H., Chang, S. H.: Tetrahedron Letters 26, 1369 (1970)
36) Kaye, H.: Macromolecules 4, 147 (1971)
37) Kaye, H., Chang, S. H.: Macromolecules 5, 397 (1971)
38) Kaye, H., Chang, S. H.: J. Makromol. Sci. A 7, 1127 (1973)
39) Kaye, H., Chou, H. J.: J. Polymer Sci. Phys. 13, 477 (1975)
40) Cassidy, F., Jones, A. S.: European Polymer J. 2, 319 (1966)
41) Boulton, M. G., Jones, A. S., Walker, R. T.: J. Chem. Soc. (C) 1968, 1216
42) Jones, A. S., Khan, M. K. A., Walker, R. T.: J. Chem. Soc. (C) 1968, 1454
43) Halford, M. H., Jones, A. S.: J. Chem. Soc. (C) 1968, 2667
44) Doel, E., Jones, A. S., Taylor, A.: Tetrahedron Letters 1963, 2285
45) Kondo, K., Murata, M., Takemoto, K.: Tech. Rep. Osaka Univ. 22, 785 (1972)

46) Takemoto, K., Kawakubo, F., Kondo, K.: Makromol. Chem. **148**, 131 (1971)
47) Okubo, T., Ban, K., Tsuji, T., Fujimoto, T., Ise, N.: submitted to Makromol. Chem.
48) Cautrecaseas, P., Anfinsen, C. B.: Methods Enzymol. **22**, 345 (1971)
49) Tüppy, H., Küchler, E.: Biochim. Biophys. Acta **80**, 669 (1964)
50) Jones, A. S., Parson, D. G.: Proc. Chem. Soc. **1961**, 78
51) Jones, A. S., Parsons, D. G., Roberts, D. G.: European Polymer J. **3**, 187 (1967)
52) Greber, G., Schott, H.: Angew. Chem. **82**, 82 (1970)
53) Schott, H., Greber, G.: Makromol. Chem. **145**, 11 (1971)
54) Schott, H., Greber, G.: Makromol. Chem. **149**, 253 (1971)
55) Schott, H., Greber, G.: Makromol. Chem. **149**, 261 (1971)
56) Tsuji, T., Okubo, T., Ise, N.: Nucleic Acids Res. Sp. No. **2**, 107 (1976)
57) Shimidzu, T., Murakami, A., Konishi, Y.: 24th Polymer Symposium, Preprint S5A04 (1975)
58) Cuatrecases, P., Wilchek, M., Anfinsen, C. B.: Proc, Natl. Acad. Sci. USA **61**, 636 (1968)
59) Dunn, B. M., Chaiken, I. M.: Proc. Natl. Acad. Sci. USA **71**, 2382 (1974)
60) Dunn, B. M., Chaiken, I. M.: Biochemistry **14**, 2343 (1975)
61) Hoffman, J. L.: Macromol. Sci. Chem. A**7**, 1147 (1973)
62) Naylor, R., Gilham, P. T.: Biochemistry **5**, 2722 (1966)
63) Sulston, J., Lohrmann, R., Orgel, L. E., Miles, H. T.: Proc. Natl. Acad. Sci. USA **60**, 409 (1968)
64) Shimidzu, T., Fukui, K.: Annual Report Res. Inst. Chem. Fiber, Japan **25**, 87 (1968)
65) Sulston, J., Lohrman, R., Orgel, L. E., Bernlowhr, H. S., Weimann, B. J., Miles, H. T.: J. Mol. Biol. **40**, 227 (1969)
66) Bernloehr, H. S., Lohrmann, R., Sulston, J., Orgel, L. E., Miles, H.: J. Mol. Biol. **47**, 257 (1970)
67) Shabarova, Z. A., Prokofiew, M. A.: FEBS Lett. **11**, 237 (1970)
68) Uesugi, S., Ts'O, P. O. P.: Biochemistry **13**, 3142 (1974)
69) Shimidzu, T., Imai, A.: Bull. Chem. Soc. Japan **49**, 349 (1976)
70) Shimidzu, T.: Private communication
71) Wetmur, J. G., Davidson, N.: J. Mol. Biol. **31**, 349 (1968)
72) Wetmur, J. G.: Biopolymers **10**, 601 (1971)
73) Lee, C. H., Wetmur, J. G.: Biopolymers **11**, 549 (1972)
74) Lee, C. H., Wetmur, J. G.: Biopolymers **11**, 1485 (1972)
75) Wetmur, J. G.: Biopolymers **14**, 2517 (1975)
76) Ross, P. D., Sturtevant, J. M.: Proc. Natl. Acad. Sci. U. S. **46**, 1360 (1960)
77) Ross, P. D., Sturtevant, J. M.: J. Amer. Chem. Soc. **84**, 4503 (1962)
78) Inman, R. B., Baldwin, R. L.: J. Mol. Biol. **8**, 452 (1964)
79) Studier, F. W.: J. Mol. Biol. **41**, 199 (1969)
80) Manning, G. S.: Biopolymers **15**, 1333 (1976)
81) Steinhardt, J. S., Fugitt, C. H.: J. Res. Natl. Bur. Standards **29**, 315 (1942)
82) Lawrence, L., Moore, W. J.: J. Amer. Chem. Soc. **73**, 3973 (1951)
83) Kern, W., Herold, W., Scherhag, B.: Makromol. Chem. **17**, 231 (1955)
84) Kern, W., Scherhag, B.: Makromol. Chem. **20**, 209 (1958)
85) Yoshikawa, S., Kim, O. K.: Bull. Chem. Soc. Japan **39**, 1515 (1966)
86) Yoshikawa, S., Kim, O. K.: Bull. Chem. Soc. Japan **39**, 1729 (1966)
87) Sakurada, I., Sakaguchi, Y., Ono, T., Ueda, T.: Makromol. Chem. **91**, 243 (1966)
88) Sakurada, I., Sakaguchi, Y., Ohmura, Y.: Bull. Inst. Chem. Res. Kyoto Univ. **44**, 135 (1966)
89) Painter, T. J., Morgan, W. T. J.: Ind. (London) **1961**, 437
90) Painter, T. J.: J. Chem. Soc. **1962**, 3932
91) Arai, K., Ise, N.: Makromol. Chem. **176**, 37 (1975)
92) Arai, K., Hagiwara, N., Ise, N.: Nippon Kagaku Kaishi **1975**, 201
93) Suzuki, K., Taniguchi, Y., Sugimoto, D., Inoue, O.: 21st Polymer Symposium, Preprints p. 1045 (1972)
94) Arcus, C. L., Gonzales, C. G., Linnecar, D. F. C.: Chem. Comm. **1969**, 1377
95) Fernández-Prini, R., Baumgartner, E.: J. Amer. Chem. Soc. **96**, 4489 (1974)

96) Okubo, T., Ise, N.: J. Org. Chem. **38**, 3120 (1973)
97) Sakurada, I., Sakaguchi, Y., Ono, T., Ueda, T.: Kobunshi, Kagaku, **22**, 696 (1965)
98) Suzuki, K., Taniguchi, K.: Presented at 21st Annual Meeting of Soc. Polymer Sci. Japan (1972)
99) Rudolfo, T., Hamilton, J. A., Cordes, E. H.: J. Org. Chem. **39**, 2281 (1974)
100) Romsted, L. R., Cordes, E. H.: J. Amer. Chem. Soc. **90**, 4404 (1968)
101) Behme, M. T. A., Fullington, J. G., Noel, R., Cordes, E. H.: J. Amer. Chem. Soc. **87**, 266 (1965)
102) Kitano, H., Okubo, T.: J. Chem. Soc. Perkin II, **1976**, 1074
103) Bunton, C. A., Robinson, L.: Amer. Chem. Soc. **90**, 5972 (1968)
104) Okubo, T., Ise, N.: Presented at 22nd Annual Meeting of Soc. Polymer Sci. Japan, Kyoto (1973)
105) Ishiwatari, T., Okubo, T., Ise, N.: Presented at 24th Annual Meeting Soc. Polymer Sci. Japan, Tokyo (1975)
106) Overberger, C. G., Salamone, J. C., Yaroslavsky, S.: Pure Appl. Chem. **15**, 453 (1967)
107) Kunitake, T., Okahata, Y.: Adv. Polymer Sci. **20**, 159 (1976)
108) Overberger, C. G., Sitaramainah, R., St. Pierre, T., Yaroslavsky, S.: J. Amer. Chem. Sco. **87**, 3270 (1965)
109) Overberger, C. G., Maki, H.: Macromolecules **3**, 214, 220 (1970)
110) Overberger, C. G., Podsiadly, C. J.: Bioorganic Chem. **3**, 35 (1974)
111) Shimidzu, T., Furuta, A., Nakamoto, Y.: Macromolecules **7**, 160 (1974)
112) Shimidzu, T., Furuta, A., Watanabe, T., Kato, S.: Makromol. Chem. **175**, 119 (1974)
113) Kunitake, T., Shinkai, S.: Makromol. Chem. **151**, 127 (1972)
114) Morawetz, H., Overberger, C. G., Salamone, J. C., Yaroslavsky, S.: J. Amer. Chem. Soc. **90**, 651 (1968)
115) Shinkai, S., Kunitake, T.: Polymer J. **4**, 253 (1973)
116) Overberger, C. G., Pacansky, J.: J. Polymer Sci. Symp. No. **45**, 39 (1974)
117) Kirsh, Y. E., Kabanov, V. A.: Kokl. Akad. Nauk, USSR, **195**, 109 (1970)
118) Ise, N., Okubo, T., Kitano, H., Kunugi, S.: J. Amer. Chem. Soc. **97**, 2882 (1975)
119) Klotz, I. M., Royer, G. P., Scarpa, I. A.: Proc. Natl. Acad. Sci. USA **68**, 263 (1971)
120) Kiefer, H. C., Congdon, W. I., Scarpa, I. S., Klotz, I. M.: Proc. Natl. Acad. Sci. USA **69**, 2155 (1972)
121) Cruickshank, P., Sheehan, J. C.: J. Amer. Chem. Soc. **86**, 2070 (1964)
122) Letsinger, R. L., Savereide, T. J.: J. Amer. Chem. Soc. **84**, 114, 3122 (1962)
123) Starodubtzev, S. G., Kirsh, Yu. E., Kabanov, V. A.: European Polymer J. **10**, 739 (1974)
124) Cho, J. R., Morawetz, H.: Macromolecules **6**, 628 (1973)
125) Klotz, I. M., Stryker, V. H.: J. Amer. Chem. Soc. **90**, 2717 (1968)
126) Royer, G. P., Klotz, I. M.: J. Amer. Chem. Soc. **91**, 5885 (1969)
127) Bink, Y., Klotz, I. M.: Bioorg. Chem. **1**, 275 (1971)
128) Pshezhetskii, V. S., Murtazaeva, G. A., Kabanov, V. A.: European Polymer J. **10**, 571 (1974)
129) a. Suzuki, S., Higashiyama, T., Nakahara, A.: Bioorg. Chem. **2**, 145 (1973)
b. Suzuki, S., Nakahara, A.: Bioorg. Chem. **4**, 250 (1975)
130) Kunitake, T., Shinkai, S., Hirotsu, S.: J. Polymer Sci. Letters ed. **13**, 377 (1975)
131) Kirsh, Y. E., Lebedeva, T. A., Kabanov, V. A.: J. Polymer Sci. Letters **13**, 207 (1975)
132) Takeishi, M., Niino, S., Hayama, S.: Makromol. Chem. **177**, 1225 (1976)
133) Winstein, S., Lucas, H. J.: J. Amer. Chem. Soc. **60**, 836 (1938)
134) Nozawa, T., Akimoto, Y., Hatano, M.: Makromol. Chem. **158**, 21 (1972)
135) Nozaqa, T., Akimoto, Y., Hatano, M.: Makromol. Chem. **161**, 289 (1972)
136) McGarvey, F. X., Kunin, R.: Ion exchange technology. Chap. 11. Nachod, F. C. and Schubert, J. (eds.). New York: Academic Press 1956
137) Underwood, G. E., Deatherage, F. E.: Science **115**, 95 (1952)
138) Paulson, J. C., Deathrage, F. E.: J. Biol. Chem. **205**, 909 (1953)
139) Whitaker, J. R., Deatherage, F. E.: J. Amer. Chem. Soc. **77**, 3360 (1955)
140) Riesz, P., Hammet, L. P.: J. Amer. Chem. Soc. **76**, 992 (1958)
141) Chen, C. H., Hammet, L. P.: J. Amer. Chem. Soc. **80**, 1329 (1958)

[142] Painter, T. J., Morgan, W. T. J.: Nature **191**, 39 (1961)
[143] Collins, R. F.: Chem. Ind. (London) **1957**, 736
[144] Hartler, N., Hyllengren, K.: J. Polymer Sci. **55**, 779 (1961)
[145] Sakurada, I., Sakaguchi, Y., Ono, T.: Bull. Inst. Chem. Res., Kyoto Univ. **43**, 149 (1965)
[146] Sakurada, I., Sakaguchi, Y., Ono, T.: Kobunshi Kagaku **24**, 570 (1967)
[147] Morita, H., Nishiyama, Y., Kitano, H., Okubo, T., Ise, N.: to be published
[148] Haskell, V. C., Hammet, L. P.: J. Amer. Chem. Soc. **71**, 1284 (1949)
[149] Affrossman, S.: J. Chem. Soc. (B) **1966**, 1015
[150] Trueblood, K. N., Lucas, H. J.: J. Amer. Chem. Soc. **74**, 1338 (1952)
[151] Keefer, R. M., Andrews, L. J., Kepner, R. E.: J. Amer. Chem. Soc. **71**, 2381 (1949)
[152] Murachi, T., Okumura, K.: J. Polymer Sci. Letters **14**, 361 (1976)
[153] Arai, K., Ogiwara, Y.: J. Appl. Polymer Sci. **20**, 1989 (1976)
[154] Morawetz, H., Vogel, B.: J. Am
[154] Morawetz, H., Vogel, B.: J. Amer. Chem. Soc. **91**, 563 (1969)
[155] Morawetz, H., Gordimer, G.: J. Amer. Chem. Soc. **92**, 7532 (1970)
[156] Ise, N., Matsui, F.: J. Amer. Chem. Soc. **90**, 4242 (1968)
[157] Brönsted, J. N.: Z. Physik. Chem. **102**, 69 (1922); **115**, 337 (1925)
[158] Bjerrum, N.: Z. Physik. Chem. **108**, 82 (1924); **118**, 251 (1925)
[159] Mita, K., Kunugi, S., Okubo, T., Ise, N.: J. Chem. Soc. Faraday I **71**, 936 (1975)
[160] Mita, K., Okubo, T., Ise, N.: J. Chem. Soc. Faraday I **72**, 1033 (1976)
[161] Shikata, M., Kim, S., Mita, K., Ise, N., Kunugi, S.: Proc. Roy. Soc. (London), **351**, 233 (1976)
[162] Okubo, T., Ise, N.: Proc. Roy. Soc. (London) A **327**, 413 (1972)
[163] Ise, N., Matsuda, Y.: J. Chem. Soc. Faraday I **69**, 99 (1973)
[164] Duystee, E. F. J., Grunwald, E.: J. Amer. Chem. Soc. **81**, 4540 (1959)
[165] Winter, L. J., Grunwald, E.: J. Amer. Chem. Soc. **87**, 4608 (1965)
[166] Albrizzio, J., Archila, J., Rodulfo, T., Cordes, E. H.: J. Org. Chem. **37**, 871 (1972)
[167] Okubo, T., Ise, N.: J. Amer. Chem. Soc. **95**, 2293 (1973)
[168] Ise, N., Shimidzu, T., Okubo, T.: Enzyme-simulated organic Reactions I (in Japanese). Chap. 4. Tokyo: Nankodo Co. (1976)

Received December 13, 1976
S. Okamura (editor)

Author Index Volumes 1–25

Allegra, G. and *Bassi, I. W.:* Isomorphism in Synthetic Macromolecular Systems. Vol. 6, pp. 549–574.
Ayrey, G.: The Use of Isotopes in Polymer Analysis. Vol. 6, pp. 128–148.
Baldwin, R. L.: Sedimentation of High Polymers. Vol. 1, pp. 451–511.
Basedow, A. M. and *Ebert, K.:* Ultrasonic Degradation of Polymers in Solution. Vol. 22, pp. 83–148.
Batz, H.-G.: Polymeric Drugs. Vol. 23, pp. 25–53.
Bergsma, F. and *Kruissink, Ch. A.:* Ion-Exchange Membranes. Vol. 2, pp. 307–362.
Berry, G. C. and *Fox, T. G.:* The Viscosity of Polymers and Their Concentrated Solutions. Vol. 5, pp. 261–357.
Bevington, J. C.: Isotopic Methods in Polymer Chemistry. Vol. 2, pp. 1–17.
Bird, R. B., Warner, Jr., H. R., and *Evans, D. C.:* Kinetik Theory and Rheology of Dumbbell Suspensions with Brownian Motion. Vol. 8, pp. 1–90.
Böhm, L. L., Chmelir̆, M., Löhr, G., Schmitt, B. J. und *Schulz, G. V.:* Zustände und Reaktionen des Carbanions bei der anionischen Polymerisation des Styrols. Vol. 9, pp. 1–45.
Bovey, F. A. and *Tiers, G. V. D.:* The High Resolution Nuclear Magnetic Resonance Spectroscopy of Polymers. Vol. 3, pp. 139–195.
Braun, J.-M. and *Guillet, J. E.:* Study of Polymers by Inverse Gas Chromatography. Vol. 21, pp. 107–145.
Breitenbach, J. W., Olaj, O. F. und *Sommer, F.:* Polymerisationsanregung durch Elektrolyse. Vol. 9, pp. 47–227.
Bresler, S. E. and *Kazbekov, E. N.:* Macroradical Reactivity Studied by Electron Spin Resonance. Vol. 3, pp. 688–711.
Bywater, S.: Polymerization Initiated by Lithium and Its Compounds. Vol. 4, pp. 66 to 110.
Carrick, W. L.: The Mechanism of Olefin Polymerization by Ziegler-Natta Catalysts. Vol. 12, pp. 65–86.
Casale, A. and *Porter, R. S.:* Mechanical Synthesis of Block and Graft Copolymers. Vol. 17, pp. 1–71
Cerf, R.: La dynamique des solutions de macromolecules dans un champ de vitesses. Vol. 1, pp. 382–450.
Cicchetti, O.: Mechanisms of Oxidative Photodegradation and of UV Stabilization of Polyolefins. Vol. 7, pp. 70–112.
Clark, D. T.: ESCA Applied to Polymers. Vol. 24, pp. 125–188.
Coleman, Jr., L. E. and *Meinhardt, N. A.:* Polymerization Reactions of Vinyl Ketones. Vol. 1, pp. 159–179.
Crescenzi, V.: Some Recent Studies of Polyelectrolyte Solutions. Vol. 5, pp. 358–386.
Davydov, B. E. and *Krentsel, B. A.:* Progress in the Chemistry of Polyconjugated Systems. Vol. 25, pp. 1–46.
Dole, M.: Calorimetric Studies of States and Transitions in Solid High Polymers. Vol. 2, pp. 221–274.
Dreyfuss, P. and *Dreyfuss, M. P.:* Polytetrahydrofuran. Vol. 4, pp. 528–590.
Dušek, K. and *Prins, W.:* Structure and Elasticity of Non-Crystalline Polymer Networks. Vol. 6, pp. 1–102.
Eastham, A. M.: Some Aspects of the Polymerization of Cyclic Ethers. Vol. 2, pp. 18–50.

Ehrlich, P. and *Mortimer, G. A.:* Fundamentals of the Free-Radical Polymerization of Ethylene. Vol. 7, pp. 386–448.
Eisenberg, A.: Ionic Forces in Polymers. Vol. 5, pp. 59–112.
Elias, H.-G., Bareiss, R. und *Watterson, J. G.:* Mittelwerte des Molekulargewichts und anderer Eigenschaften. Vol. 11, pp. 111–204.
Fischer, H.: Freie Radikale während der Polymerisation, nachgewiesen und identifiziert durch Elektronenspinresonanz. Vol. 5, pp. 463–530.
Fujita, H.: Diffusion in Polymer-Diluent Systems. Vol. 3, pp. 1–47.
Funke, W.: Über die Strukturaufklärung vernetzter Makromoleküle, insbesondere vernetzter Polyesterharze, mit chemischen Methoden. Vol. 4, pp. 157–235.
Gal'braikh, L. S. and *Rogovin, Z. A.:* Chemical Transformations of Cellulose. Vol. 14, pp. 87–130.
Gandini, A.: The Behaviour of Furan Derivatives in Polymerization Reactions. Vol. 25, pp. 47–96.
Gerrens, H.: Kinetik der Emulsionspolymerisation. Vol. 1, pp. 234–328.
Goethals, E. J.: The Formation of Cyclic Oligomers in the Cationic Polymerization of Heterocycles. Vol. 23, pp. 103–130.
Graessley, W. W.: The Etanglement Concept in Polymer Rheology. Vol. 16, pp. 1 to 179.
Hay, A. S.: Aromatic Polyethers. Vol. 4, pp. 496–527.
Hayakawa, R. and *Wada, Y.:* Piezoelectricity and Related Properties of Polymer Films. Vol. 11, pp. 1–55.
Heitz, W.: Polymeric Reagents. Polymer Design, Scope, and Limitations. Vol. 23, pp. 1–23.
Helfferich, F.: Ionenaustausch. Vol. 1, pp. 329–381.
Hendra, P. J.: Laser-Raman Spectra of Polymers. Vol. 6, pp. 151–169.
Henrici-Olivé, G. und *Olivé, S.:* Kettenübertragung bei der radikalischen Polymerisation. Vol. 2, pp. 496–577.
Henrici-Olivé, G. und *Olivé, S.:* Koordinative Polymerisation an löslichen Übergangsmetall-Katalysatoren. Vol. 6, pp. 421–472.
Henrici-Olivé, G. and *Olivé, S.:* Oligomerization of Ethylene with Soluble Transition-Metal Catalysts. Vol. 15, pp. 1–30.
Hermans, Jr., J., Lohr, D., and *Ferro, D.:* Treatment of the Folding and Unfolding of Protein Molecules in Solution According to a Lattic Model. Vol. 9, pp. 229 to 283.
Hutchison, J. and *Ledwith, A.:* Photoinitiation of Vinyl Polymerization by Aromatic Carbonyl Compounds. Vol. 14, pp. 49–86.
Iizuka, E.: Properties of Liquid Crystals of Polypeptides : with Stress on the Electromagnetic Orientation. Vol. 20, pp. 79–107.
Imanishi, Y.: Syntheses, Conformation, and Reactions of Cyclic Peptides. Vol. 20, pp. 1–77.
Inagaki, H.: Polymer Separation and Characterization by Thin-Layer Chromatography. Vol. 24, pp. 189–237.
Inoue, S.: Asymmetric Reactions of Synthetic Polypeptides. Vol. 21, pp. 77–106.
Ise, N.: Polymerizations under an Electric Field. Vol. 6, pp. 347–376.
Ise, N.: The Mean Activity Coefficient of Polyelectrolytes in Aqueous Solutions and Its Related Properties. Vol. 7, pp. 536–593.
Isihara, A.: Intramolecular Statistics of a Flexible Chain Molecule. Vol. 7, pp. 449 to 476.
Isihara, A.: Irreversible Processes in Solutions of Chain Polymers. Vol. 5, pp. 531 to 567.
Isihara, A. and *Guth, E.:* Theory of Dilute Macromolecular Solutions. Vol. 5, pp. 233–260.
Janeschitz-Kriegl, H.: Flow Birefringence of Elastico-Viscous Polymer Systems. Vol. 6, pp. 170–318.
Jenngins, B. R.: Electro-Optic Methods for Characterizing Macromolecules in Dilute Solution. Vol. 22, pp. 61–81.
Kawabata, S. and *Kawai, H.:* Strain Energy Density Functions of Rubber Vulcanizates from Biaxial Extension. Vol. 24, pp. 89–124.
Kennedy, J. P. and *Chou, T.:* Poly(isobutylene-*co*-β-Pinene): A New Sulfur Vulcanizable, Ozone Resistant Elastomer by Cationic Isomerization Copolymerization. Vol. 21, pp. 1–39.
Kennedy, J. P. and *Gillham, J. K.:* Cationic Polymerization of Olefins with Alkylaluminium Initators. Vol. 10, pp. 1–33.

Kennedy, J. P. and *Johnston, J. E.:* The Cationic Isomerization Polymerization of 3-Methyl-1-butene and 4-Methyl-1-pentene. Vol. 19, pp. 57–95.

Kennedy, J. P. and *Langer, Jr., A. W.:* Recent Advances in Cationic Polymerization. Vol. 3, pp. 508–580.

Kennedy, J. P. and *Otsu, T.:* Polymerization with Isomerization of Monomer Preceding Propagation. Vol. 7, pp. 369–385.

Kennedy, J. P. and *Rengachary, S.:* Correlation Between Cationic Model and Polymerization Reactions of Olefins. Vol. 14, pp. 1–48.

Kissin, Yu. V.: Structures of Copolymers of High Olefins. Vol. 15, pp. 91–155.

Kitagawa, T. and *Miyazawa, T.:* Neutron Scattering and Normal Vibrations of Polymers. Vol. 9, pp. 335–414.

Knappe, W.: Wärmeleitung in Polymeren. Vol. 7, pp. 477–535.

Koningsveld, R.: Preparative and Analytical Aspects of Polymer Fractionation. Vol. 7, pp. 1–69.

Kovacs, A. J.: Transition vitreuse dans les polymers amorphes. Etude phénoménologique. Vol. 3, pp. 394–507.

Krässig, H. A.: Graft Co-Polymerization of Cellulose and Its Derivatives. Vol. 4, pp. 111–156.

Kraus, G.: Reinforcement of Elastomers by Carbon Black. Vol. 8, pp. 155–237.

Krimm, S.: Infrared Spectra of High Polymers. Vol. 2, pp. 51–72.

Kuhn, W., Ramel, A., Walters, D. H., Ebner, G. and *Kuhn, H. J.:* The Production of Mechanical Energy from Different Forms of Chemical Energy with Homogeneous and Cross-Striated High Polymer Systems. Vol. 1, pp. 540–592.

Kunitake, T. and *Okahata, Y.:* Catalytic Hydrolysis by Synthetic Polymers. Vol. 20, pp.159–221.

Kurata, M. and *Stockmayer, W. H.:* Intrinsic Viscosities and Unperturbed Dimensions of Long Chain Molecules. Vol. 3, pp. 196–312.

Ledwith, A. and *Sherrington, D. C.:* Stable Organic Cation Salts: Ion Pair Equilibria and Use in Cationic Polymerization. Vol. 19, pp. 1–56.

Lee, C.-D. S. and *Daly, W. H.:* Mercaptan-Containing Polymers. Vol. 15, pp. 61–90.

Lipatov, Y. S.: Relaxation and Viscoelastic Properties of Heterogeneous Polymeric Compositions. Vol. 22, pp. 1–59.

Mano, E. B. and *Coutinho, F. M. B.:* Grafting on Polyamides. Vol. 19, pp. 97–116.

Meyerhoff, G.: Die viscosimetrische Molekulargewichtsbestimmung von Polymeren. Vol. 3, pp. 59–105.

Millich, F.: Rigid Rods and the Characterization of Polyisocyanides. Vol. 19, pp. 117–141.

Morawetz, H.: Specific Ion Binding by Polyelectrolytes. Vol. 1, pp. 1–34.

Mulvaney, J. E., Oversberger, C. C., and *Schiller, A. M.:* Anionic Polymerization. Vol. 3, pp. 106–138.

Okubo, T. and *Ise, N.:* Synthetic Polyelectrolytes as Models of Nucleic Acids and Esterases. Vol. 25, pp. 135–181.

Osaki, K.: Viscoelastic Properties of Dilute Polymer Solutions. Vol. 12, pp. 1–64.

Oster, G. and *Nishijima, Y.:* Fluorescence Methods in Polymer Science. Vol. 3, pp. 313–331.

Overberger, C. G. and *Moore, J. A.:* Ladder Polymers. Vol. 7, pp. 113–150.

Patat, F., Killmann, E. und *Schiebener, C.:* Die Absorption von Makromolekülen aus Lösung. Vol. 3, pp. 332–393.

Peticolas, W.L.: Inelastic Laser Light Scattering from Biological and Synthetic Polymers. Vol. 9, pp. 285–333.

Pino, P.: Optically Active Addition Polymers. Vol. 4, pp. 393–456.

Plesch, P. H.: The Propagation Rate-Constants in Cationic Polymerisations. Vol. 8, pp. 137–154.

Porod, G.: Anwendung und Ergebnisse der Röntgenkleinwinkelstreuung in festen Hochpolymeren. Vol. 2, pp. 363–400.

Postelnek, W., Coleman, L. E., and *Lovelace, A. M.:* Fluorine-Containing Polymers. I. Fluorinated Vinyl Polymers with Functional Groups, Condensation Polymers, and Styrene Polymers. Vol. 1, pp. 75–113.

Rogovin, Z. A. and *Gabrielyan, G. A.:* Chemical Modifications of Fibre Forming Polymers and Copolymers of Acrylonitrile. Vol. 25, pp. 97–134.

Roha, M.: Ionic Factors in Steric Control. Vol. 4, pp. 353–392.
Roha, M.: The Chemistry of Coordinate Polymerization of Dienes. Vol. 1, pp. 512 to 539.
Safford, G. J. and *Naumann, A. W.:* Low Frequency Motions in Polymers as Measured by Neutron Inelastic Scattering. Vol. 5, pp. 1–27.
Schuerch, C.: The Chemical Synthesis and Properties of Polysaccharides of Biomedical Interest. Vol. 10, pp. 173–194.
Schulz, R. C. und *Kaiser, E.:* Synthese und Eigenschaften von optisch aktiven Polymeren. Vol. 4, pp. 236–315.
Seanor, D. A.: Charge Transfer in Polymers. Vol. 4, pp. 317–352.
Seidl, J., Malinský, J., Dušek, K. und *Heitz, W.:* Makroporöse Styrol-Divinylbenzol-Copolymere und ihre Verwendung in der Chromatographie und zur Darstellung von Ionenaustauschern. Vol. 5, pp. 113–213.
Semjonow, V.: Schmelzviskositäten hochpolymerer Stoffe. Vol. 5, pp. 387–450.
Semlyen, J. A.: Ring-Chain Equilibria and the Conformations of Polymer Chains. Vol. 21, pp. 41–75.
Sharkey, W. H.: Polymerizations Through the Carbon-Sulphur Double Bond. Vol. 17, pp. 73–103.
Shimidzu, T.: Cooperative Actions in the Nucleophile-Containing Polymers. Vol. 23, pp. 55–102.
Slichter, W. P.: The Study of High Polymers by Nuclear Magnetic Resonance. Vol. 1, pp. 35–74.
Small, P. A.: Long-Chain Branching in Polymers. Vol. 18, pp. 1–64.
Smets, G.: Block and Graft Copolymers. Vol. 2, pp. 173–220.
Sohma, J. and *Sakaguchi, M.:* ESR Studies on Polymer Radicals Produced by Mechanical Destruction and Their Reactivity. Vol. 20, pp. 109–158.
Sotobayashi, H. und *Springer, J.:* Oligomere in verdünnten Lösungen. Vol. 6, pp. 473–548.
Sperati, C. A. and *Starkweather, Jr., H. W.:* Fluorine-Containing Polymers. II. Polytetrafluoroethylene. Vol. 2, pp. 465–495.
Sprung, M. M.: Recent Progress in Silicone Chemistry. I. Hydrolysis of Reactive Silane Intermediates. Vol. 2, pp. 442–464.
Stille, J. K.: Diels-Alder Polymerization. Vol. 3, pp. 48–58.
Szwarc, M.: Termination of Anionic Polymerization. Vol. 2, pp. 275–306.
Szwarc, M.: The Kinetics and Mechanism of N-carboxy-α-amino-acid Anhydride (NCA) Polymerization to Poly-amino Acids. Vol. 4, pp. 1–65.
Szwarc, M.: Thermodynamics of Polymerization with Special Emphasis on Living Polymers. Vol. 4, pp. 457–495.
Tani, H.: Stereospecific Polymerization of Aldehydes and Epoxides. Vol. 11, pp. 57–110.
Tate, B. E.: Polymerization of Itaconic Acid and Derivatives. Vol. 5, pp. 214–232.
Tazuke, S.: Photosensitized Charge Transfer Polymerization. Vol. 6, pp. 321–346.
Teramoto, A. and *Fujita, H.:* Conformation-dependent Properties of Synthetic Polypeptides in the Helix-Coil Transition Region. Vol. 18, pp. 65–149.
Thomas, W. M.: Mechanism of Acrylonitrile Polymerization. Vol. 2, pp. 401–441.
Tobolsky, A. V. and *DuPré, D. B.:* Macromolecular Relaxation in the Damped Torsional Oscillator and Statistical Segment Models. Vol. 6, pp. 103–127.
Tosi, C. and *Ciampelli, F.:* Applications of Infrared Spectroscopy to Ethylene-Propylene Copolymers. Vol. 12, pp. 87–130.
Tosi, C.: Sequence Distribution in Copolymers: Numerical Tables. Vol. 5, pp. 451 to 462.
Tsuchida, E. and *Nishide, H.:* Polymer-Metal Complexes and Their Catalytic Activity. Vol. 24, pp. 1–87.
Tsuji, K.: ESR Study of Photodegradation of Polymers. Vol. 12, pp. 131–190.
Valvassori, A. and *Sartori, G.:* Present Status of the Multicomponent Copolymerization Theory. Vol. 5, pp. 28–58.
Voorn, M. J.: Phase Separation in Polymer Solutions. Vol. 1, pp. 192–233.
Werber, F. X.: Polymerization of Olefins on Supported Catalysts. Vol. 1, pp. 180 to 191.
Wichterle, O., Šebenda, J., and *Králiček, J.:* The Anionic Polymerization of Caprolactam. Vol. 2, pp. 578–595.
Wilkes, G. L.: The Measurement of Molecular Orientation in Polymeric Solids. Vol. 8, pp. 91–136.

Wöhrle, D.: Polymere aus Nitrilen. Vol. 10, pp. 35–107.
Wolf, B. A.: Zur Thermodynamik der enthalpisch und der entropisch bedingten Entmischung von Polymerlösungen. Vol. 10, pp. 109–171.
Woodward, A. E. and *Sauer, J. A.:* The Dynamic Mechanical Properties of High Polymers at Low Temperatures. Vol. 1, pp. 114–158.
Wunderlich, B. and *Baur, H.:* Heat Capacities of Linear High Polymers. Vol. 7, pp. 151–368.
Wunderlich, B.: Crystallization During Polymerization. Vol. 5, pp. 568–619.
Wrasidlo, W.: Thermal Analysis of Polymers. Vol. 13, pp. 1–99.
Yamazaki, N.: Electrolytically Initiated Polymerization. Vol. 6, pp. 377–400.
Yoshida, H. and *Hayashi, K.:* Initiation Process of Radiation-induced Ionic Polymerization as Studied by Electron Spin Resonance. Vol. 6, pp. 401–420.
Zachmann, H. G.: Das Kristallisations- und Schmelzverhalten hochpolymerer Stoffe. Vol. 3, pp. 581–687.
Zambelli, A. and *Tosi, C.:* Stereochemistry of Propylene Polymerization. Vol. 15, pp. 31–60.

B. Vollmert

Polymer Chemistry

Translated from the German by E. H. Immergut
630 figures. XVII, 652 pages. 1973
ISBN 3-540-05631-9

This book gives a comprehensive coverage of the synthesis of polymers and their reactions, structure, and properties. The treatment of the reactions used in the preparation of macromolecules and in their transformation into crosslinked materials is particularly detailed and complete. The book also gives an up-to-date presentation of other important topics, such as enzymatic and protein synthesis, solution properties of macromolecules, polymer crystallization, and properties of polymers in the solid state.

The content and presentation of Professor Vollmert's book is more encompassing than most existing treatises, and its numerous figures and tables convey a wealth of data, never, however, at the expense of intellectual clarity or educational value.

The presentation is mainly on a fundamental and general level and yet the reader – student or professional – is gradually and almost casually introduced to all important natural and synthetic polymers. Complicated phenomena are explained with the aid of the simplest available examples and models in order to ensure complete understanding. However, the reader is also encouraged to think for himself and even to criticize the author's point of view.

All of the chapters have been revised and enlarged from the German edition, and many of the sections are entirely new.

Springer-Verlag
Berlin Heidelberg New York

B. Rånby, J. F. Rabek

ESR Spectroscopy in Polymer Research

356 figures, 32 tables. Approx. 450 pages. 1977
(Polymers/Properties and Applications, Volume 1)
ISBN 3-540-08151-8

The main purpose of this book is to collect the present available information on the applications of electron spin resonance (ESR) spectroscopy in polymer research. The book has been written both for those who want an introduction to this field, and for those who are already familiar with ESR and are interested in application to polymers. Therefore, the fundamental principles of ESR spectroscopy are first outlined, the experimental methods including computer applications are described in more detail, and the main emphasis is on the application of ESR methods to polymer problems. The authors hope that this book will provide a useful source of information by giving a coherent treatment and extensive references to original papers, reviews, and discussions in monographs and books. In this way we hope to encourage polymer chemists, organic chemists, biochemists, physicists, and material scientists to apply ESR methods to their research problems. (2519 references).

Springer-Verlag
Berlin Heidelberg New York

RETURN CHE[CK]
TO → ... Hall   642-3753